Project Report 40　　　　　　　　　　　　　　　　　　　　　London, 2000

Grassmoor Lagoons: organic sludge bioremediation field trials

Summary

This report is a case study of a field trial for the *ex-situ* remediation of coal tar sludges from disposal lagoons at Grassmoor, Derbyshire, prior to the redevelopment of this contaminated site. The report describes the nature of the contamination found at the site and also the potential hazards resulting from the contamination.

The unusual nature of the type and scale of contamination necessitated the need to assess a number of remediation techniques. The proposed technique of bioremediation meant that the usual procurement methods would not meet the trial's objectives of minimising the financial and technical risks to both the client and treatment contractor. The report therefore includes sections on the selection procedures used for appointing the field trials contractor and also the contractual framework used.

Additional laboratory testing and monitoring enabled independent verification of the treatment process and an assessment to be made of residual contamination likely to remain at the end of the treatment programme. Further sections of the report draw on the outcomes of the field trials to offer a financial assessment of the treatment process and present conclusions drawn from the field trials and the remediation methods used.

Grassmoor Lagoons: organic sludge bioremediation field trials

Taylor, M J, Storey, P and Westcott, F J

Construction Industry Research and Information Association

Project Report 40   © CIRIA 2000   ISBN 0 86017 840 4

Keywords
Contaminated land, field trials, bioremediation, bench trials, technology demonstrations, procurement, polyaromatic, hydrocarbon, organic sludge

Reader interest	Classification	
Geotechnical, civil, petrochemical and environmental engineers; developers; regulators	AVAILABILITY	Unrestricted
	CONTENT	Record of field and laboratory
	STATUS	Committee-guided
	USER	Geotechnical, environmental and development professionals; regulators

Published by CIRIA, 6 Storey's Gate, Westminster, London SW1P 3AU. All rights reserved. No part of this publication may be reproduced or transmitted in any form or by any other means, including photocopying and recording, without the written permission of the copyright holder, application for which should be addressed to the publisher. Such written permission must also be obtained before any part of this publication is stored in a retrieval system of any nature.

Acknowledgements

This report, a case study of the Grassmoor Lagoons bioremediation field trial, forms part of Research Project 489 under the third phase of CIRIA's research programme on the remedial treatment of contaminated land. The objective of this programme is to promote safe, effective and economic remediation using the most appropriate technologies under UK conditions. The specific objective was to report on, and draw guidance from, a case study involving field trials of biological remediation of lagoon sludges contaminated with mineral oils and polyaromatic hydrocarbons (PAHs).

The report was written under contract to CIRIA by M J Taylor, P Storey and F J Westcott of Derbyshire County Council.

Following established CIRIA practice, the research was guided by a steering group, which comprised:

Mr B Hardisty (chair)	English Partnerships
Dr P Bardos	R^3 Environmental Technology Ltd.
Dr J Birnstingl	British Aerospace Defence Ltd.
Prof N Christofi	Napier University
Dr P Daley	British Gas R and T
Mr E Gray	English Partnerships
Ms J Hodgson	Environment Agency
Dr M Lambson	BP International Ltd.
Dr G Lethbridge	Shell Research
Dr P Morgan	ICI Engineering Technology
Mr M Morton	Environment Agency
Mr S Redfearn	The BOC Foundation
Mr N Smith	British Waterways
Mr J Spence Watson	Confederation of Landscape Industries in Scotland

Corresponding members were:

Mr R Berry	Department of the Environment, Transport and the Regions
Dr A Hart	Environment Agency
Mr P Wilson	Highways Agency

CIRIA's research managers for the project were Dr M R Sansom and F M Jardine.

CIRIA is grateful to Derbyshire County Council, the client for the trial, for permission to use the material contained within this report, and for the support and technical input of Derbyshire Consulting Engineers that made this study possible.

CIRIA also gratefully acknowledges the following organisations for funding Phase III of CIRIA's programme "The remedial treatment of contaminated land":

- Department of the Environment, Transport and the Regions, Construction Directorate
- The BOC Foundation
- British Waterways
- The Environment Agency
- Highways Agency
- National House-Building Council
- Scottish Enterprise.

The authors would like to acknowledge the following for their involvement in the project: R Blaney, Dr B Ellis and Dr J Rees of CELTIC Technologies Ltd.

Executive summary

1. This report describes a case study of field trials for the remediation of a heavily contaminated site, prior to its redevelopment as an extension to an existing country park. The strategic objectives of the reclamation scheme are to return the site to public use by remediating the contamination while protecting the on-site water environment and preventing off-site migration of contaminants. The field trials focused on the treatment on site of concentrated, industrial waste organic sludges.

2. The history and previous uses of the site which resulted in the contamination are described, along with the findings of earlier site investigations. Previous studies have identified the nature and quantities of contaminated material, broadly categorised as water, sludges and colliery spoil. The case study specifically reports on the nature of the heavily contaminated sludges and discusses several options for their remediation. The remediation approaches considered comprised both off-site and on-site solutions. Of these approaches, the *ex-situ* biological treatment method offered the potential for the best environmental and most economic solution.

3. The unusual and innovative nature of the project and proposed action meant that there were relatively high degrees of financial and technical risk for both the client and the treatment contractor. In order to minimise these risks and to ensure that the adopted approach was effective, Derbyshire County Council considered it prudent to undertake laboratory and field-scale trials prior to commencing full-scale works. Independent chemical testing to verify the results of the treatment process was also undertaken.

4. As a result of discussions between Derbyshire County Council and CIRIA about the unusual aspects of the project, the County Council were commissioned by CIRIA to undertake the reporting of the trials as a case study under CIRIA's demonstration programme. Funding from CIRIA enabled further independent chemical testing to be carried out to assist in the verification process.

5. Treatment-based remediation methods require specialist knowledge and competencies. Derbyshire County Council therefore determined that a specialist soil treatment contractor should be appointed. This report sets out the selection criteria and strategy used for identifying and procuring the services of such a contractor. Use was made of questionnaires, focusing on financial stability, health and safety records, quality assurance and project-specific requirements. Celtic Technologies Ltd was the appointed contractor.

6. The roles of the various organisations involved in the trials were as indicated below.

Role	Organisation
Client	Derbyshire County Council
Funding agency	English Partnerships
Scheme designer and project manager	Derbyshire Consulting Engineers
Contractor for the trials	Celtic Technologies Limited
Contractor's chemical testing sub-contractor	Robertson Laboratories
Client's independent chemical testing sub-contractor	Alcontrol UK Limited

7. The results of the bench trials were used as a basis for the field-scale trials, which commenced with the incremental addition of sludge and treatment additives to a specially prepared treatment bed and the subsequent construction of a biopile. The phased mixing of the material, combined with the parallel system of treatment bed and biopile, were used to overcome difficulties created by inclement weather and the very wet nature of the sludge.

8. Extensive monitoring of the treatment material throughout all stages of the treatment processes as well as continued environmental monitoring in and around the site was undertaken. This was done not only to verify the outcomes of the trials, but also to verify that health and safety and environmental requirements were met in full.

9. Mean concentrations of key organic contaminant indicators measured in the treatment bed and the biopile over the progress of the trials were as follows.

Contaminants	Treatment bed (mg/kg)		Biopile (mg/kg)	
Time	PAH (EPA 16)	Mineral oils (soil)	PAH (EPA 16)	Mineral oils (soil)
After second sludge addition (21.2.98)	11324	8824	11210	10031
At end of trials (14.4.98)	3355	7213	7082	11183
26 weeks after second sludge addition (end August 1998)	1104		4736	
71 weeks after second sludge addition (end October 1999)	474	1859	540	2785

10. Mean concentrations of key inorganic contaminant indicators measured in the treatment bed and the biopile over the progress of the trials were as follows.

Contaminants	Treatment bed (mg/kg)			Biopile (mg/kg)		
Time	Mercury	Lead	Arsenic	Mercury	Lead	Arsenic
After second sludge addition (21.2.98)	1.9	58	85	2.2	65	99
At end of trials (14.4.98)	1.6	70	113	1.9	74	82
16 weeks after second sludge addition (12.6.98)	1.6	63	101			

These, being largely unchanged, are indications that the reductions in organic contamination were not achieved by dilution.

11. Pointers for improved practice that stem from the study are:
    - to use controls with bench trials, when there is the possibility of volatilisation of the constituents of the contamination, eg naphthalene
    - to consider how to deal with the imprecision of the sampling and analytical procedures (particularly a problem with PAHs in soil)
    - to examine the possibility of optimising degradation through further laboratory studies on the effectiveness of nutrient additions, use of surfactants, etc.

12. The trials, despite being carried out in difficult conditions, demonstrate that the heavily contaminated sludge material present at the Grassmoor site is amenable to remediation by *ex-situ* biological means in practical field conditions and that the outcomes are economically feasible.

Contents

Summary 2
Acknowledgements 3
Executive summary 5
List of figures 11
List of tables 12
List of boxes 14
Glossary 15
Abbreviations 16

1 Reclamation and case study objectives and strategy 17
    1.1 Reclamation scheme objectives 17
    1.2 Field trial and case study objectives 18
    1.3 CIRIA programme of technology demonstrations 20

2 The Grassmoor Lagoons reclamation scheme 21
    2.1 Introduction 21
    2.2 Description and history of the site 21
    2.3 Geology and mining 26
    2.4 Hydrology and hydrogeology 27
    2.5 Ground investigation 28
    2.6 Contamination of soil, sludge and water 31
    2.7 Sources, pathways and receptors 32
    2.8 Proposed remedial action 33
    2.9 Context for the trials 35

3 Management of the the treatment trials 37
    3.1 Evaluation of treatment options 37
    3.2 Project management 37
    3.3 Selection process for the treatment contractor 38
    3.4 Procurement for the treatment trials 40
    3.5 Result of selection procedure 44

4 The treatment trials 45
    4.1 Programme of works 45
    4.2 Further sludge characterisation 45
    4.3 Bench trials 46
    4.4 Mechanical handling trials 47
    4.5 Field trials 48
    4.6 Environmental monitoring 51
    4.7 Health and safety 51
    4.8 Public information 52

5	Results of the trials	53
	5.1 Sludge characterisation	53
	5.2 Bench trial	55
	5.3 Treatment bed/biopile during sludge addition	56
	5.4 Treatment bed/biopile after sludge addition	58
	5.5 Treatment bed/biopile at end of trial period	59
	5.6 Longer-term monitoring	62
	5.7 Residual concentrations related to earthworks disposition and risk-based end-points	64
6	Cost analysis	65
	6.1 Original cost estimate for the trials	65
	6.2 Costs on completion of the trial	65
7	Outcomes of the trials	67
	7.1 Procurement	67
	7.2 Treatment method	67
	7.3 Sludge characterisation	68
	7.4 Environmental monitoring	68
	7.5 Field trials	69
8	Conclusions from the case study	71
	8.1 Management process	71
	8.2 Recommendations for future work	71
	8.3 Trials programme	73

Appendices ... 75

A1 Management process flowchart for sludge treatment as originally proposed .. 75
A2 Selection questionnaire ... 78
A3 Summary of the sludge characterisation results ... 84
A4 Summary of environmental monitoring results ... 85
A5 Information sheet ... 87
A6 Summary of results of sampling the bench trials ... 88
A7 Summary of results of sampling treatment bed (independent analysis) ... 94
A8 Summary of results of treatment bed monitoring ... 105
A9 Summary of results of sampling treatment bed after completion of the trials ... 112

References ... 119

LIST OF FIGURES

2.1	Plan of the Grassmoor Lagoons site	22
2.2	Lagoon A from the south tip, 1995	23
2.3	Layout of the site, 1947	24
2.4	Schematic section through the site	25
2.5	Method of effluent treatment, 1986–1993	26
2.6	Over-water ground investigation of Lagoon C, 1995	29
3.1	Flowchart for the initial screening of questionnaire responses	39
3.2	Structure of documents for the treatment trial procurement	42
4.1	Excavation of sludge in Lagoon A	48
4.2	The original treatment bed	49
4.3	Rotovation of the treatment bed	50
4.4	Layout of the biopile	51
5.1	Degradation of DRH and PAH content during the bench trials	55
5.2	Layout of the original treatment bed	56
5.3	Layout of the revised treatment bed	58
5.4	The progress of degradation	60
5.5	Degradation of PAH continuing after the trials	64
A1.1	Management process flowchart for sludge treatment, as originally proposed	76
A6.1	Concentrations of selected contaminants in Treatment T1	91
A6.2	Concentrations of selected contaminants in Treatment T3	92
A6.3	Concentrations of selected contaminants in Treatment T6	92
A6.4	Concentrations of total PAHs in Treatment T3 and Treatment T6	93
A8.1	Average VOCs recorded in Treatment bed 1	106
A8.2	Average VOCs recorded in Treatment bed 2	106
A8.3	Average VOCs recorded in Treatment bed 3	108
A8.4	Average carbon dioxide levels recorded in the stockpile	109
A8.5	Average VOCs measured in the stockpile	111

LIST OF TABLES

2.1	Volumes of contaminated material inferred from the site investigation	31
2.2	Possible source-pathway-receptor hazard scenarios	33
3.1	Assessment matrix for the tenders for the trials	43
4.1	Composition of the bench-scale treatment microcosms	47
5.1	Definition of EPA 16 and WSR 6 PAHs	54
5.2	Characterisation of lagoon sludges	54
5.3	Initial contamination concentrations	57
5.4	Concentrations after second sludge addition	59
5.5	Concentrations at end of treatment trials	61
5.6	Concentrations in the original treatment bed and temporary stockpile at end of trials	61
5.7	Concentrations in treatment bed 16 weeks after second sludge addition	63
5.8	PAH concentrations in treatment bed and biopile 26 weeks after second sludge addition	63
5.9	Concentrations in the treatment bed and biopile 71 weeks after second sludge addition	65
6.1	Original costings for the activity schedule	66
6.2	Events leading to increased costs	
A3.1	Chemical contaminant testing summary – sludge characterisation	84
A4.1	Results of personal monitoring for VOCs	85
A4.2	VOC monitoring at site boundaries	85
A6.1	Summary of chemical testing of selected organic contaminants in lagoon sludge and colliery spoil	88
A6.2	Summary of chemical testing of selected volatile organic contaminants in lagoon sludge	89
A6.3	Summary of chemical testing of inorganic contaminants in lagoon sludge and colliery spoil	89
A6.4	Concentrations of selected contaminants – Treatment T1	90
A6.5	Concentrations of selected contaminants – Treatment T3	90
A6.6	Concentrations of selected contaminants – Treatment T6	91
A6.7	Summary of microcosm treatment parameters	93
A7.1	Chemical contaminant testing summary – treatment beds and constituents before sludge addition	94
A7.2	Chemical contaminant testing summary – original treatment bed after first sludge addition	95
A7.3	Chemical contaminant testing summary – treatment bed/biopile before second sludge addition	95
A7.4	Chemical contaminant testing summary – treatment bed after second sludge addition	96
A7.5	Chemical contaminant testing summary – biopile after second sludge addition	97

A7.6	Chemical contaminant testing summary (leach test samples) – treatment bed/biopile after second sludge addition	98
A7.7	Chemical contaminant testing summary – treatment bed four weeks after second sludge addition	99
A7.8	Chemical contaminant testing summary (leach test samples) – treatment bed four weeks after second sludge addition	99
A7.9	Chemical contaminant testing summary – treatment bed seven weeks after second sludge addition	100
A7.10	Chemical contaminant testing summary (leach test results) – treatment bed leach test results	101
A7.11	Chemical contaminant testing summary – biopile seven weeks after second sludge addition	101
A7.12	Chemical contaminant testing summary (leach test sampling) – biopile leachate test results	102
A7.13	Chemical contaminant testing summary – original treatment bed at end of trials	102
A7.14	Chemical contaminant testing summary – temporary stockpile remainder at end of trials	103
A7.15	Chemical contaminant testing summary (leachate samples) – leachate test results: old treatment bed and stockpile remainder	103
A7.16	Chemical contaminant testing summary – water sample test results	104
A8.1	Bed 1 monitoring: total concentration of VOCs (ppm)	105
A8.2	Bed 1 monitoring: pH	105
A8.3	Bed 1 monitoring: moisture content (percentage)	105
A8.4	Bed 2 monitoring: total concentration of VOCs (ppm)	107
A8.5	Bed 3 monitoring: total concentration of VOCs (ppm)	108
A8.6	Bed monitoring: pH	109
A8.7	Bed monitoring: moisture content (percentage)	109
A8.8	Stockpile monitoring: carbon dioxide (percentage volume in air)	110
A8.9	Stockpile monitoring: total concentration of VOCs (ppm)	111
A9.1	Chemical contaminant testing summary (leach test samples) – treatment bed 16 weeks after second sludge addition and 9 weeks after end of trials	112
A9.2	Chemical contaminant testing summary – treatment bed 16 weeks after second sludge addition and 9 weeks after end of trials	113
A9.3	Chemical contaminant testing summary – treatment bed 26 weeks after second sludge addition and 19 weeks after end of trials	114
A9.4	Chemical contaminant testing summary – biopile 26 weeks after second sludge addition and 19 weeks after end of trials treatment	115
A9.5	Chemical contaminant testing summary – treatment bed 71 weeks after second sludge addition and 64 weeks after end of trials	116
A9.6	Chemical contaminant testing summary – biopile 71 weeks after second sludge addition and 64 weeks after end of trials	117

LIST OF BOXES

1.1	Strategic objectives for Grassmoor Lagoons reclamation scheme	17
1.2	Field trial objectives	19
2.1	Objectives of the site investigation	28
2.2	Scope of the site investigation fieldwork	30
3.1	Project management roles	38
3.2	Key attributes of the treatment trial procurement	41
4.1	Additives and their purposes	46
4.2	Plant used in the field trials	48
5.1	Changes to contaminant mass during the field trials	62

Glossary

Air sparging	Introduction of air under pressure into the ground.
Biological treatment	Degradation of organic contaminants to harmless end-products.
Biopile	A heap of contaminated material and admixtures for the purpose of allowing bioremediation.
Bioremediation	The use of biological processes to transform or alter the structure of organic contaminants.
Colliery spoil	Rock wastes brought to the surface during the mining operation and/or material separated from coal during coal processing operations.
Hydrocarbon	One of a very large group of chemical compounds composed only of carbon and hydrogen.
Incineration	The application of heat to contaminated material either to destroy contaminants or render them less harmful.
Lagoon base sludges	Oily and tarry sludges located at the base of lagoons, originating from the settling out of coking plant effluent.
Microcosms	Small-scale mixtures of contaminated material and admixtures to represent field conditions under controlled laboratory conditions.
Public procurement procedure	Formal procedure used by central and local government bodies in the purchase of goods and services.
Solidification	Solidifying contaminated material by binding them with an insoluble matrix offering low leaching characteristics.
Thermal desorption	Heating of contaminated material to evaporate and remove volatile contaminants.
Treatment bed	Prepared level area of ground receiving contaminated material and admixtures for subsequent bioremediation treatment.

Abbreviations

AOD	above Ordnance Datum
BSI	British Standards Institution
CIRIA	Construction Industry Research and Information Association
COD	chemical oxygen demand
DCC	Derbyshire County Council
DCE	Derbyshire Consulting Engineers
DRH	diesel-range hydrocarbon
EC	European Community
ISO	International Standards Organisation
NCB	National Coal Board
NSF	National Smokeless Fuels
PAH	polyaromatic hydrocarbon
PID	photo-ionisation detector
PPE	personal protective equipment
UK	United Kingdom
US EPA	United States of America: Environmental Protection Agency
VOC	volatile organic compound
WSR	The Water Supply (Water Quality) Regulations 1989

1 Reclamation and case study objectives and strategy

1.1 RECLAMATION SCHEME OBJECTIVES

The case study describes an element of the reclamation works, comprising the bio-remediation of organic sludges carried out at the Grassmoor Lagoons site. These works were a part of the reclamation scheme for Grassmoor Lagoons. The scheme is part of the Derbyshire County Council Land Reclamation Programme, formerly funded through Derelict Land Grant (McCafferty, 1993) and now funded by English Partnerships.

The objectives of the reclamation scheme for Grassmoor Lagoons, set out in Box 1.1, are three-fold:

- to return the currently fenced lagoons area from which the public are excluded to public amenity use
- to protect the water environment at the site
- to prevent contaminants from migrating off land owned by Derbyshire County Council.

Box 1.1   *Strategic objectives for Grassmoor Lagoons reclamation scheme*

Use of country park

It is intended to return the currently fenced area from which the public are excluded to public amenity use and reclamation will be to a standard appropriate to the site's end use as a landscaped amenity area.

The reclamation scheme will remove sources of unpleasant odours, which lower the amenity of the country park and the surrounding communities.

Protection of the water environment

The water environment at the site will be controlled to minimise leaching of contaminants and to ensure that contaminated water is intercepted and, where necessary, treated before entering the River Rother or its tributaries.

The groundwater level in the perched water table will be kept as low as practicable to minimise downwards seepage into the solid Coal Measures strata and the old mine-workings under the site.

Drainage measures and capping will ensure that surface water and precipitation is kept separate from contaminated water and is kept free of contamination to ensure that the surface waters are not polluted.

Prevention of off-site migration

The reclamation scheme will aim to prevent contaminants from migrating off land owned by Derbyshire County Council.

Technical feedback

Monitoring, testing and analysis will be carried out as part of the reclamation scheme. This feedback will demonstrate its function and enable knowledge gained in executing the scheme to be documented and made available to assist in design of further phases of the scheme.

The proposed end use is thus "soft" and relatively non-intensive. The lagoons site is situated in an area between the Grassmoor Country Park, created by Derbyshire Council in 1973 from a reclaimed colliery site, and a golf course constructed on restored opencast land. Thus the area forms a green wedge of open land given over primarily to amenity use, and a development-led reclamation scenario for the site was regarded as neither practicable nor desirable. Without the realisation of added value from development land sale proceeds, the onus was upon Derbyshire County Council to develop an approach to reclamation which, while protecting the public in the area of the site, could be demonstrated to be cost-effective as well as environmentally sustainable. This approach is in accordance with the "suitable for use" concept – a key element of the UK contaminated land policy (Harris *et al*, 1997).

Reclamation works at the Grassmoor Lagoons site were prioritised in order to achieve the desired objectives. The initial priority was to lower the level of water in the lagoons, and in the colliery spoil around the lagoons, in order to minimise the potential for groundwater seepage-led migration of contaminants. This activity also enabled access to be gained to the lagoon bed sludges in the main lagoons. Lowering of the lagoon water was achieved by establishing a pumped discharge under consent to a local foul sewer. This element of the works was commenced in late 1996 (Westcott, 1997) and is continuing to operate at the time of writing. It does not form part of the case study that is the subject of this report.

The next priority is to deal with the seriously contaminated lagoon bed sludges, and the final priority will be to regrade the lagoons area to a landform satisfactory both from the visual aspect and surface water drainage considerations and, where the colliery spoil material in the tips was severely contaminated, to cap these areas with an impermeable capping material.

The second of these priorities, dealing with the seriously contaminated lagoon base sludges, is the subject of the trials described in the current case study. Dealing with the lagoon base sludges is a prerequisite for the continuation of further reclamation work, involving the creation of a new landform and surface drainage system on the site.

1.2 FIELD TRIAL AND CASE STUDY OBJECTIVES

Dealing with contaminated sludges and soils will require the major remediation effort at this site. The sludges are difficult materials to handle; they are semi-liquid, highly contaminated with coal tars and other hydrocarbons, and extremely malodorous. Possible generic remediation approaches (Martin and Bardos, 1995; USEPA, 1995a) considered included:

- off-site disposal
- on-site disposal in a specially formed and lined waste disposal facility
- *in-situ* containment
- on-site treatment and retention.

Of these generic approaches, the *ex-situ* biological treatment approach appeared to have the potential to offer the best environmental and most economic solution. Although bioremediation of organic compounds, including PAH, has previously been carried out and reported (USEPA, 1995b and 1996a), the combination of adverse physical conditions, the high concentrations of contaminants, the volume requiring remediation and the ambient climatic conditions was more onerous.

It was apparent that the effectiveness of the approach could not be assumed to be certain. It was therefore considered prudent that laboratory bench-scale treatability studies and field-scale trials should be carried out before committing the County Council to full-scale works using this approach. In designing an effective means of undertaking field-scale trials, it was necessary to identify objectives. These objectives are set out in Box 1.2.

Box 1.2   *Field trial objectives*

- To demonstrate the effectiveness of biological treatment of the sludges encountered on the Grassmoor Lagoons site.
- To reduce the financial and technical risk of the project, both to Derbyshire County Council and to the contractor.
- To enable the optimum combination of treatment method, cost, time and treatment area to be established.
- To optimise the requirements for inputs such as nutrients, heat, etc.
- To determine the nature of outputs such as solid material, waters and gaseous material, and the degree of residual contamination likely to remain.
- To enable independent verification and audit testing of the process to be carried out by testing contractors employed by Derbyshire County Council.

In addition, it was necessary that the trials should be carried out within a contractual framework to allow Derbyshire County Council to select its specialist treatment contractors through a transparent public procurement procedure, yet allow the results of the trials and the information and knowledge gained therein to be assimilated into the programme for the treatment works within this contractual framework.

The aims of the trials were thus not only to demonstrate the effectiveness of biological treatment of the sludges at the site, but also:

- to reduce the financial and technical risk of the project, both to Derbyshire County Council and to the contractor
- to enable the optimum combination of treatment method, cost, time and treatment area to be established
- to optimise the requirements for inputs such as nutrients
- to determine the nature of outputs such as solid materials, liquids and gaseous materials, and the degree of residual contamination likely to remain at the end of the treatment programme.

The trials were to be the subject of independent verification and audit testing of the process, which would be carried out by testing contractors employed directly by Derbyshire County Council.

Derbyshire County Council was aware of the unusual and innovative nature of the project and had been in communication with CIRIA. As a result of these discussions, the County Council was commissioned by CIRIA in 1997 to undertake the reporting of these trials under the CIRIA case study demonstration programme.

This publication describes the background to the trials set in the context of the reclamation scheme as a whole. It includes an outline of the desk study and ground investigation works, the interpretation of the ground conditions and the execution and analytical interpretation of the field trials.

The case study is reported not only in terms of the degree of efficacy of the technology trialled and factors such as practical operational and cost issues, but also those of procurement and integration of the technology into a broader reclamation programme for the site.

1.3 CIRIA PROGRAMME OF TECHNOLOGY DEMONSTRATIONS

CIRIA's contaminated land research programme is intended to provide guidance for dealing with contaminated land. The case study demonstration programme consists of a series of projects, reported under commission to CIRIA and in accordance with their reporting requirements (Harris, 1996), which form the third phase of the contaminated land research programme. The objective of this collaborative research is to promote safe, effective and economical remediation through the promotion of research and development, and the dissemination of guidance and information.

A specific objective is to provide detailed information through site application reports, which cover a number of potentially viable and widely applicable remedial treatment technologies for contaminated soil and groundwater. This report presents a case study of the application of biological treatment of contaminated sludges by use of enhanced treatment beds and biopiles, and is based on the field-scale treatment trials carried out on behalf of Derbyshire County Council as a part of the Grassmoor Lagoons reclamation scheme.

2 The Grassmoor Lagoons reclamation scheme

2.1 INTRODUCTION

The decline of the deep-mined coal industry left a legacy of dereliction and contamination in the former coalfields of north-eastern Derbyshire. As well as the highly visible scars left by the spoil tips of the old collieries, the industries associated with the processing and use of coal have also left their own – often less visible but more insidious – damage to the environment. Of these industries, perhaps the most damaging from a contamination and pollution point of view was the coal carbonisation or coking industry. Coal carbonisation involves the purification of coal by pyrolysis and, as well as the main products of coke and coal gas, a range of often polluting and toxic by-products are created. Some of these by-products have commercial use as feedstocks for chemical industry, but many have no commercial value and are disposed of as solid or liquid wastes (Environmental Resources Ltd, 1987).

In response to increased demand, due to the expansion of the steel industry and the designation of smokeless zones in UK towns and cities, the coking industry was modernised and expanded from the mid-1950s onwards. One of the major investments made in the industry during this period was the construction of the Avenue coking plant, near Wingerworth, Chesterfield (Pegg and Ashworth, 1986). This plant opened in the late 1950s and produced smokeless fuel from then until its closure in 1993.

The treatment of liquid effluent waste from the Avenue coking plant was not carried out within the confines of the plant itself, but at some distance from it. Effluent from the plant was pumped by pipeline approximately 3 kilometres to the colliery at Grassmoor, where it was discharged into lagoons formed within the spoil tips of the colliery. It underwent basic pre-treatment, before being discharged to the foul sewer.

The collieries, the coking plants and the associated infrastructure of mineral railways and materials handling facilities have now all been abandoned, but the legacy of pollution and contamination remains a challenge to the regeneration of these former coalfield areas. Since the late 1960s, Derbyshire County Council has actively pursued the reclamation of derelict land in the North Derbyshire coalfield, with the dual purposes of providing sites for new industries to replace the employment lost from the coal industry, and to improve the physical environment of the area as an aid to this regenerative process. The Grassmoor Lagoons scheme forms part of this programme. Its reclamation intended to complete the regeneration of the area of the five collieries formerly situated between Chesterfield and Clay Cross.

2.2 DESCRIPTION AND HISTORY OF THE SITE

The site of the Grassmoor Lagoons is located between the villages of Grassmoor, Corbriggs and Temple Normanton, some four miles south-east of Chesterfield. The site is situated within a green wedge of land, a "pay and play" golf course established on a restored opencast coal site bounding it to the west, and the Derbyshire County Council's Grassmoor Country Park bounding it to the north, east and south.

The country park was established by Derbyshire County Council on the reclaimed site of the former Grassmoor colliery and coking plant in the mid-1970s. When the site was bought from the National Coal Board (NCB), their subsidiary, National Smokeless Fuels (NSF), held a lease on around 10 hectares of the old colliery tips. This area was used for the treatment and disposal of effluent in a series of lagoons, within an area fenced off from the country park. This use was continued until early 1993, when the lease on the lagoons areas was terminated by Derbyshire County Council.

The lagoons area, which is fenced to prevent public access, includes the highest ground in the country park with panoramic views over Chesterfield and the Rother Valley. It is shown in Figure 2.1. The high ground is formed from the former colliery spoil tips designated for convenience the "south tip" and the "north tip". As shown in Figure 2.2, the area is separated by deep depressions in which the low level lagoons A and C are situated.

Two sets of smaller lagoons are present on top of the former spoil tips: one group (P, F, G and H and the infilled D and E) on the south tip, and the other group (I, J, K, O and N) on the north tip (see Figure 2.1). Lagoon L is situated at the base of the spoil tip, to the north of the north tip, acting as a balancing pond for the foul sewer discharge.

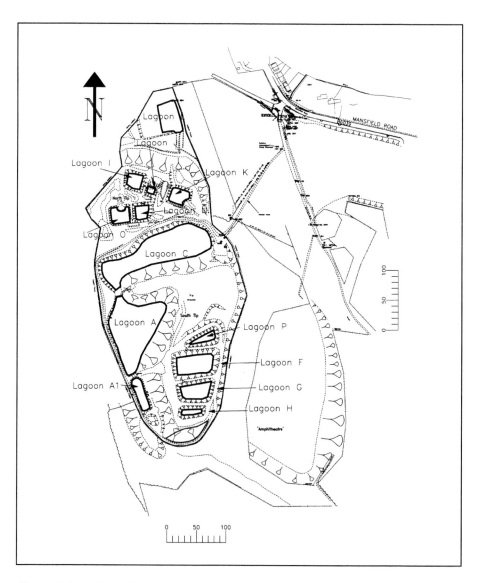

Figure 2.1   *Plan of the Grassmoor Lagoons site*

Figure 2.2 *Lagoon A from the south tip, 1995*

The lagoons area has not been subject to landscaping, but there is natural regeneration of vegetation on some of the less contaminated spoil tip slopes. Snakes, foxes and ground-nesting birds are known to inhabit the site, despite its polluted state.

The history of the Grassmoor Lagoons site is, in microcosm, the history of the North Derbyshire coalfield. In 1840 the surrounding area was predominantly pastoral. The area now occupied by the Grassmoor Lagoons site, and that of the former colliery, is shown to have been two valleys with small northerly-flowing streams, the larger to the east called the Calow Brook and a smaller unnamed stream to the west, which joined adjacent to the present-day Corbriggs gate to the country park.

The development of this area into the North Derbyshire coalfield commenced with the coming of the railways, the Midland Railway reaching Chesterfield in 1840. George Stephenson, the engineer to the railway, recognised the mineral wealth in the coal measures in the area and established the Clay Cross Company to exploit these reserves of coal and iron. One of the large collieries established during this period was the Grassmoor Colliery. Its development was made possible by the establishment of a network of branch railway lines, linking the area to the main line railways.

The Number 1 shaft for the Grassmoor Colliery was sunk in 1861. By 1880 the colliery was well developed alongside the Calow Brook. Development of the colliery continued throughout the latter part of the nineteenth century, and the area taken up by spoil tipping is shown on contemporary mapping to have expanded steadily to the west of the colliery. By 1918, further development, including the construction of the Grassmoor coking plant to the south of the colliery, had been carried out and areas of the tips are shown on contemporary mapping as being given over to filter beds. By 1938 the spoil tipping to the west of the Grassmoor Colliery had expanded further to the west, necessitating the diversion of the western stream around the toe of the spoil tips. The diverted alignment of the western stream follows the base of the present-day low level lagoons A and C.

After the end of the Second World War, and with the nationalisation of the coal industry in 1947, further expansion and development occurred at Grassmoor, as shown in Figure 2.3. This development included the expansion of the spoil tipping area to the north and west of the diverted western stream course. The pre-1938 tipping area to the south and east of the stream course, now disused for tipping, was graded to a flat top and appears to have been in use as a lagoon area. Further lagoons were established on the newer tips during the early 1950s. During the late 1950s the western stream was finally diverted into a piped culvert, discharging directly into the Calow Brook, and the former stream course to the south of the deep lagoons was backfilled with colliery spoil.

Grassmoor Colliery was virtually exhausted of economically winnable reserves of coal by the late 1960s, and was closed in 1967. Meanwhile, the spoil tipping area had been taken over for use by the Avenue plant for effluent treatment. Initially, the system consisted of the two low-level lagoons, A and C, and the high-level lagoons on the southern tip, B, D, E, F, G and H. In about 1986, the system was extended by the construction of the higher level lagoons on the north tip, I, J, K, O and N. This included the provision of a forced aeration system using a system of helixer aerators to improve the quality of the effluent by simple air stripping prior to its discharge to the foul sewer. Lagoons D and E were abandoned and infilled.

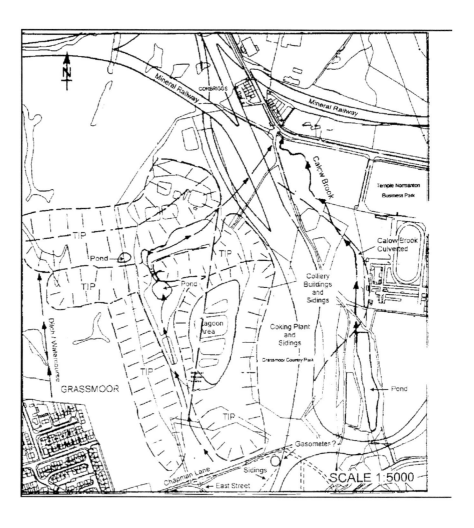

Figure 2.3 *Layout of the site, 1947*

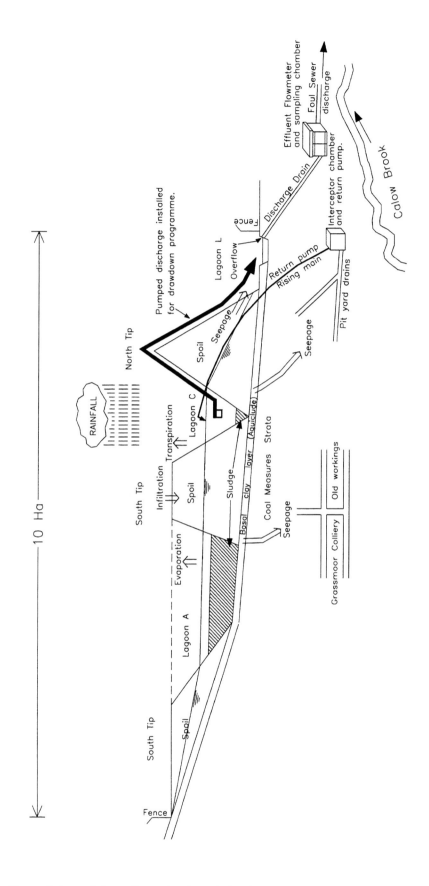

Figure 2.4 *Schematic section through the site*

The lagoons system, as extended in 1986, remained in use until the closure of the site in 1993 and is shown diagrammatically in Figures 2.4 and 2.5. Hence the contaminative activities taking place on this site had been occurring for several decades before the site was taken over by Derbyshire County Council. By far the greatest impact, however, has been created by the Avenue coking plant during the period 1960–1993.

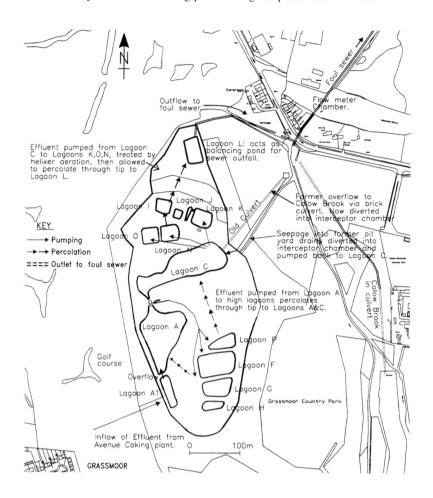

Figure 2.5   *Method of effluent treatment, 1986–1993*

2.3   GEOLOGY AND MINING

The site is situated on the outcrop of the Westphalian series, formerly referred to as the Coal Measures, of the Carboniferous period.

The measures encountered below the site were formerly known as the Middle Coal Measures, and the site is situated slightly to the east of the westernmost outcrop of the Top Hard coal seam in the North Derbyshire coalfield. The coal measures generally dip towards the east in this coalfield, but in the Grassmoor area the strata are folded into a synclinal basin trending north-west to south-east, named the Williamthorpe Syncline. To the east of this trough lies an anticlinal structure designated the Brimington Anticline, also trending north-west to south-east. The presence of the syncline has the effect of pushing the Top Hard outcrop westwards and created a favourable location to establish a colliery. Extraction plans of the colliery indicate that under the site most major seams, as far down as the Blackshale seam 450 m below ground, have been extracted (Natural Environment Research Council, 1967).

The rock faces of the coal measures in the area of the site are predominantly mudstones and siltstones, but former NCB geological plans indicate two particular sandstone bands within the solid strata overlying the Top Hard seam which were encountered during the site investigation. Although the mudstone would be expected to have low–medium permeability, the effect of mining subsidence on these rocks has increased their fracture porosity so that measured permeabilities are similar to those of the sandstones.

Significant drift deposits were not encountered in the area of the site and are generally absent in this area. The mudstones, at outcrop, have generally weathered to give rise to a thin layer (1–3 m thick) of eluvial or colluvial clayey superficial soil with low permeability, which tends to form an aquiclude.

The most significant determinant of ground conditions in the lagoons area is the made ground, which consists of a mixture of colliery spoil from the deep mine workings dumped over a period of around a century. Early spoil is unlikely to have been washed and may contain significant proportions of coal fines. Certainly there is evidence, both anecdotal and from visual appearance of materials, that the spoil in the tips has been burned, either deliberately or through spontaneous combustion. The permeability of this material is generally medium to high, which contrasts with the low permeability of the weathered mudstones lying immediately below.

2.4 HYDROLOGY AND HYDROGEOLOGY

Of the natural watercourses, only the partially culverted Calow Brook remains, running from south to north through two ponds originally impounded to provide water for the colliery. Formerly, a western tributary stream ran north through what is now the lagoons site before joining the Calow Brook at Corbriggs. Its course was periodically moved to the west to allow the colliery tips to expand during the early part of this century until it was finally diverted into a concrete culvert running directly to the Calow Brook, to the south of the lagoons site.

The former colliery had a network of surface water drains which discharged into the Calow Brook. Many of these drains remain in the ground, although they have been covered by restoration fill placed during the reclamation of the colliery site to form the country park in the early 1970s.

These drains were found to be intercepting contaminated seepage from the former lagoons site, allowing it to discharge into the brook. As part of the extensions to the lagoons system constructed in 1986 by NSF, the drains were intercepted at a chamber, into which an automatic float actuated submersible pump was placed, which pumped the effluent back into the lower level lagoon C. Surface water falling on the country park area is intercepted by a series of surface drains and is led directly to the Calow Brook.

Where groundwater strikes within the solid strata were encountered during the site investigation, they were sporadic and ephemeral, and it is apparent that the solid coal measures strata are still underdrained by the disused mine workings below the site. The mine workings in the area of the Williamthorpe Syncline are drained into the disused Williamthorpe Colliery, where pumps operated by the NCB (now the Coal Authority) continue to be operated to prevent the migration of the water down dip and through the network of disused mine workings into the deep Nottinghamshire mines which continue in operation.

The undulating topography of the lagoons area and the presence of the lagoons cause all surface precipitation that falls on the lagoons site to remain in the area, unless it is able to migrate away from the area by seepage through the ground. The permeability of the colliery spoil allows this precipitation to infiltrate readily into the spoil, where it forms a perched water table, the basal aquiclude consisting of the weathered mudstones which are situated at or immediately below the original ground profile, buried under the colliery spoil.

Seepage from this perched water table occurs in two ways; through the basal aquiclude and thence into the deep mine workings, or sub-horizontally, through the colliery spoil and the restoration fill for the reclamation scheme, and thence into the former colliery yard drains, which are intercepted and their discharges returned to lagoon C. Some seepage occurs at the northern margin of the north tip, which discharges into lagoon L, where it overflows to the foul sewer. The elements of this complex hydrological system are illustrated in Figure 2.4. The hydrology of the site has been changed during the reclamation scheme by the installation and operation of the lagoons drawdown system, which is described briefly in Section 2.8 below.

2.5 GROUND INVESTIGATION

Ground investigation was carried out on the Grassmoor Lagoons site between March and August 1995. The main investigation was intended to provide a broad overall coverage of the site sufficient to establish the general subsoil conditions and degree of contamination present. It was accepted that, dependent on the selected reclamation methodology, it might be necessary to carry out some supplementary investigation at detailed design stage. The objectives of the site investigation are summarised in Box 2.1.

Box 2.1 *Objectives of the site investigation*

---

1. To profile thicknesses and depths of made ground and superficial strata.

2. To establish the nature and depth of the solid coal measures strata.

3. To confirm whether the Top Hard coal seam underlying the site had been extracted and whether any open voids remain.

4. To determine the presence, depth and direction of movement of water in the ground and the behaviour of seepage from the lagoons.

5. To determine geotechnical properties of the made ground, the subsoil strata and the sludges in the base of the lagoons.

6. To determine the nature, extent and concentration of contaminants in the colliery spoil tips and the lagoon sludges.

7. To determine the nature, extent and concentration of contaminants in the groundwater and the surface water in the lagoons.

8. To determine the depth of water and thickness of the sludges in the lagoons.

9. To confirm the alignment of the brick culvert running from Lagoon C to the Corbriggs gate and the former effluent pipeline from Avenue Coking Plant.

10. To install groundwater monitoring standpipes within the site and boundary monitoring standpipes at the periphery of the site and determine the baseline groundwater conditions at these monitoring points.

At the time of the ground investigation at the lagoons site, both the low-level lagoons A and C remained full of effluent. There was no way of drawing these lagoons down prior to the investigation. It was therefore determined that the most effective means of investigating the ground conditions below the lagoons was by a combination of conventional cable-tool percussion boreholes from pontoons floating on the lagoons, dynamic probe profiles, also driven from pontoons, and echo soundings carried out from a dinghy, to ascertain the bed profile of the lagoons (see Figure 2.6).

Figure 2.6   *Over-water ground investigation of Lagoon C, 1995*

The echo soundings in lagoon A revealed that a shallow depth of water was present, overlying soft material. Following preliminary water quality tests to confirm that there was no hazard associated with pumping water between the low-level lagoons A and C, sufficient depth of water to enable the pontoon to float on lagoon A was achieved by means of pumping from lagoon C.

Where possible, permeability tests were carried out, either by means of variable head permeability testing in boreholes, or by tests carried out on U100 open-drive tube samples from the boreholes. Other than the use of pontoon-mounted percussion rigs and dynamic probe equipment, the investigation was carried out using conventional site investigation methods.

Box 2.2 shows the general scope of the ground investigation, which confirmed that a silty clay layer generally 1–3 m thick was present underneath the colliery spoil throughout the lagoons area. This layer followed the profile of the original ground, and was interpreted as weathered mudstone from coal measures strata, of eluvial or colluvial origin. Below this layer, the coal measures strata were progressively less weathered. The less weathered mudstones and siltstones of the coal measures strata showed generally high permeabilities, considered to be the result of the effects of mining subsidence (estimated to be as much as 8 m in total at the Grassmoor site).

Box 2.2   *Scope of the site investigation fieldwork*

Cable-tool percussion boreholes

- 41 in total were bored to depths between 3 m and 30 m
- five were bored over water (lagoons A and C) from floating pontoons
- 12 had permanent monitoring standpipes installed
- SPTs were carried out in all boreholes
- disturbed and (where appropriate) U100 open-drive tube samples were taken
- groundwater was sampled where encountered
- some variable head permeability testing carried out in standpipes.

Rotary drillholes

- Eight were drilled in total to depths between 4.6 m and 43 m
- six had permanent monitoring standpipes installed
- all were drilled adjacent to percussion boreholes with open hole drilling through superficial strata and continuous rotary coring in solid strata using semi-rigid plastic core liner.

Trial pits/trenches

- 90 were dug in total to depths between 2 m and 4.2 m
- five were mechanically excavated trial trenches
- two were hand-excavated trial trenches
- disturbed samples were taken
- groundwater was sampled where encountered.

Dynamic probe locations

- 12 were driven in total to depths between 4.1 m and 10 m
- apparatus conformed to DPH equipment requirements in BS 1377:1990
- probings were made over water (lagoons A and C) from floating pontoons
- probing was continued to refusal or to 10 m depth.

Depth soundings

Some 90 in total were carried out using echo-sounding equipment.

The original ground surface is in the form of a valley which runs from the south to the north of the lagoons site and, at the position of the present lagoon C, turns to the north-east, heading in the direction of the Corbriggs gate. This is in accordance with the historical data referred to in Section 2.2 above. The seepage of groundwater at the site tends to be controlled by the combination of the original ground profile in the form of a valley, with the very low permeability silty clay layer (generally the coefficient of permeability being in the region of 5×10^{-10} m/s) acting as an aquiclude at the base of the spoil tips.

Of particular significance was the fact, revealed by the investigation, that both the low-level lagoons A and C, and the small lagoon A1 situated immediately to the south of lagoon A, had significant quantities of malodorous black oily and tarry sludge in their bases. In the case of lagoon A, the thickness of this sludge was revealed as being as much as 8 m. However, due to the volume of the overlying water and the fact that, as it was itself contaminated, it could not be allowed to discharge to surface watercourses, the further investigation and characterisation of this sludge was severely restricted at the ground investigation stage.

2.6 CONTAMINATION OF SOIL, SLUDGE AND WATER

The ground investigation confirmed that the Grassmoor lagoons site was significantly contaminated as a result of its previous uses. The contamination affects water, both in the lagoons themselves and within the perched groundwater table in the spoil tips, the sludges in the base of the lagoons and the colliery spoil within the spoil tips themselves.

The major proportion of the former spoil heaps is contaminated to some degree. The southern area of spoil tipping, south of lagoon C, which has been used since the First World War for lagooning of coking plant effluents from both the Grassmoor and Avenue plants, shows the most significant contamination, concentrations of PAH recorded in it being characteristically around 10 000 mg/kg within the spoil itself, and higher values being recorded in areas where sludge appears to be present. The tipping area to the north of lagoon C shows less evidence of PAH contamination, having only been used since 1986 as part of the lagoon system, but some evidence of contamination is noted, apparently associated with seepage of water from lagoon C to lagoon L.

The most significant contamination problem on the site is related to the oily and tarry sludges which are at the base of the low-level lagoons, particularly lagoons A, A1 and C. The sludges are believed to have originated by settling out of the effluent in the lagoons, which itself is believed to have been a product of the process to refine the town gas created at the coking plant. The sludges are semi-liquid, black and extremely malodorous, and contain a significant concentration of oily and tarry organic materials, as well as solids (probably derived from lime or limestone dust) and water. The quantity of sludges inferred to be present on site is approximately 50 000 m^3, this quantity being an estimate which requires to be confirmed when the overlying water can be removed. Table 2.1 summarises the nature and volumes of the various materials present on site.

Table 2.1 *Volumes of contaminated material inferred from the site investigation*

Material	Estimated volume	Nature of contamination
Water	30 000 m^3 (lagoons) 120 000 m^3 (ground)	Ammonia, total cyanide Sulphate
Sludge	50 000 m^3	Very high concentrations of PAH, mineral oils, VOCs, sulphates, sulphides, phenols, total cyanides
Colliery spoil	800 000 m^3	PAH, sulphates, sulphides, total cyanides

In 1993, when the lagoons site was reoccupied by Derbyshire County Council, the lagoons were full of effluent, the top water level of lagoon A being approximately 113 m AOD and the water level in lagoon C being approximately 109.5 m AOD. The effluent in the lagoons remained contaminated with a variety of pollutants at the time of the ground investigation in 1995, despite the fact that no effluent had been discharged to the site since 1993. Pollutants included ammonia, phenol, sulphates, chlorides, total cyanides and, in some cases, PAHs.

Similar contamination was observed in samples of groundwater taken from the perched water table within the spoil tips during the excavation of boreholes and subsequently from monitoring standpipes installed in the boreholes.

It is clear that the water in the lagoons and the perched groundwater in the tips are in hydraulic continuity. The ground investigation and subsequent interpretation also indicated that the quality of the water in the lagoons was improving only slowly, despite the degree of dilution from surface precipitation, as most surface precipitation on the site seeped into the ground and was leaching contaminants from the colliery spoil before seeping into the lagoons.

A complex series of interactions is occurring at the Grassmoor lagoons site between surface water, groundwater, soils and sludges. The interpretation of the ground investigation results has enabled the nature of these interactions to be inferred with reasonable confidence. It is apparent that a significant degree of contamination from coal carbonisation waste is ubiquitous in all materials on the site. Infiltration of precipitation in to the colliery spoil tips also results in the leaching of sulphides and sulphates due to the presence of naturally occurring iron pyrite within the carboniferous rock represented in the colliery spoil material.

2.7 SOURCES, PATHWAYS AND RECEPTORS

Although interactions between the colliery spoil, the effluent and perched groundwater and the lagoon base sludges are complex, consideration of sources, pathways and receptors can conveniently be made on the basis of treating the colliery spoil, the lagoon base sludges and the effluent/groundwater as separate sources. The interaction between these separate sources can be represented in the source-pathway-receptor analysis by considering the sources also as receptors for the relevant pathways. For example, contaminated water in lagoons is a receptor for contamination leached from the spoil tips. Table 2.2 summarises the hazard scenarios for the three hazard sources.

It is apparent that not only the contamination, but also the physical nature of the sludges is a significant risk driver. The semi-liquid nature of the sludges confers a quicksand-like consistency which is extremely hazardous, particularly to children or others who might trespass on the site and venture on to them. While the PAH compounds which form a major constituent of the coal tars present constitute a serious risk to human receptors because of their being carcinogenic, their comparatively low solubility means that they are less significant in terms of the water environment than are ammonia and members of the cyanide group, in particular free cyanides.

Most of the risks associated with the colliery spoil and the sludges are from contact between the contaminants and human receptors or construction materials. There is, therefore – at least in the short term – some protection from the maintenance of the chain link and barbed wire fence which surrounds the site. The water-borne contaminants, however, are prone to seepage from the site. They could be released by catastrophic failure of the lagoon bunds or by their breaking into existing surface water drainage systems in the country park. It was considered likely that they were causing off-site pollution already. It was thus considered that an early priority was to reduce the magnitude of the hazard constituted by this source. This action was also supported by the need to gain access to the lagoon base sludges as a prerequisite to their remediation, which would necessitate the removal of the overlying water from within the lagoons.

Table 2.2 *Possible source-pathway-receptor hazard scenarios*

Hazard source	Risk-driving contaminants	Pathway	Receptor
Colliery spoil	PAH	Ingestion/adsorption	Trespassers, Country Park users, site operatives
	Phenol, cyanides, sulphates, sulphides	Oxidation/leaching	Effluent/perched groundwater
	Sulphates, sulphides	Contact/chemical attack	Concrete/cement structures, surface vegetation
	Methane/CO_2	Produced from organic material, migration	People in confined spaces
Sludges	PAH, other organic compounds	Ingestion/adsorption	Trespassers, Country Park users
		Oxidation/leaching	Effluent/perched groundwater
		Contact, chemical attack	Plastics
	VOCs	Volatilisation/wind transport/ inhalation	Local residents, Country Park users
	Physical nature (quicksand consistency)	Accidental drowning, asphyxiation	Trespassers, Country Park users
Effluent and perched groundwater	Ammonia, cyanide, COD	Surface flow or shallow seepage; by failure of lagoon bunds or overtopping	Local watercourses (Calow Brook, River Rother)
	PAH, heavy organic compounds	Deep seepage into mine workings, pumped at Williamthorpe	Local watercourses (Muster Brook, River Rother)
		Settlement	Lagoon base sludge
		Sorption on to colliery spoil	Colliery spoil
		Discharge consent failure	Foul sewer/treatment works

Although the reduction of the magnitude of the hazard posed by the lagoon waters was regarded as the earliest priority, the ground investigation revealed that in terms of magnitude and difficulty of the remediation task, and probably cost, the lagoon base sludges would pose the greatest challenge to the scheme.

2.8 PROPOSED REMEDIAL ACTION

Remediation options for the three hazard sources were considered separately, although it was accepted that each remediation action had the potential to impact on remediation actions related to the other hazard sources. The options for dealing with the contaminated water can be summarised under five headings.

1. Discharge to foul sewer for off-site treatment.

2. Transport by road tanker for off-site treatment.

3. Pre-treatment on site followed by discharge under consent to foul sewer.

4. Pre-treatment on site followed by transport off site by road tanker.

5. Full treatment on site followed by discharge to surface watercourse under Environment Agency consent.

By far the most economical of these options was discharge to foul sewer, although limits on the volume allowed to be discharged would mean a long timescale for this operation. After obtaining the discharge consent from Yorkshire Water Services, this first option was implemented. It commenced in October 1996 and is still continuing (Westcott, 1997). Approximately 70 000 m³ of water have been discharged to the sewer under this consent.

The volume of contaminated colliery spoil on the site was estimated to be in excess of 800 000 m³. Since most of this spoil is at a considerable depth below current ground level, it is not considered likely to come into contact with human receptors and the only viable pathway is by leaching due to surface water infiltration. The option of excavation and treatment or containment of the entire spoil tip is not considered to be economically viable, and could indeed increase the short-term risks due to the volume of material that would be excavated and brought to the surface, increasing the risk of exposure to sensitive receptors.

Proposals for dealing with the contaminated colliery spoil involve the regrading of the landform to a visually pleasing appearance. It will also be designed to shed surface water readily, assisted by the provision of an infiltration barrier which will form a cap to the spoil tip (Privett *et al*,1996). The outline of this proposal has been developed, but design detail will be dependent on the methods chosen for the remediation of the lagoon base sludges, and the volume and quality (in terms of residual contamination) of the materials arising from the treatment process.

Dealing with the contaminated sludges will require the major remediation effort at the site. Because of the semi-liquid nature of the sludges and their high levels of contamination it was not practical to predict which remediation methods would be effective and economic at the stage of interpretation of the ground investigation. The nature and levels of the contamination, and the quantity and consistency of the material, render off-site disposal an unattractive and probably highly expensive (if achievable at all) proposition. On this assumption, the remaining possibilities can be divided into three categories: on-site disposal, *in-situ* containment and on-site treatment and retention.

The particular circumstances of the Grassmoor site and the material in the lagoon base sludges militate against on-site disposal as an option. First, the concentrations of the organic contaminants are such that a highly engineered liner would be necessary, and the nature of the material is such that synthetic liners would probably have a short service life. Secondly, the semi-liquid nature of the sludges means that costly pre-treatment would be necessary to ensure their mechanical stability within a waste disposal repository.

Thirdly, and perhaps most significantly, entombment of these highly contaminated sludges would not decrease the contamination, but leave it to be dealt with at some future date, when the containment liners become unserviceable.

The alternatives to on- or off-site disposal consist of a series of possible treatment technologies. These can be considered under the generic headings (Martin and Bardos, 1995; USEPA, 1996b) of biological treatment; extraction; thermal/incineration, and solidification/immobilisation. Also considered, and investigated at one point by a third party, was the possibility that the sludges might have a value as a chemical feedstock, although these investigations did not bear fruit.

Of the treatment technologies investigated, the most promising appeared to be the biological treatment of the sludges. The semi-liquid nature of the sludges meant that a range of possible biological treatment technologies could be considered, ranging from a conventional oil industry land-farming approach to the use of a slurry-phase bioreactor that would treat the sludges in suspension in an aqueous liquid. What all of these approaches have in common is that they are *ex-situ* approaches; the physical nature of the sludges make *in-situ* remediation impractical.

Other than the use of biological treatment, the only technology that was considered to have a reasonable chance of success was the use of a thermal or incineration-based approach, and this was rejected both on grounds of cost and on grounds of the potential risk of release of airborne pollution in the event of incinerator or thermal plant malfunction. This is an issue that is of particular public sensitivity in the north-eastern part of Derbyshire, due to the airborne release of dioxins in the Bolsover area in the period prior to the early 1990s.

The drawbacks of the biological treatment options include the fact that it is a new technology, and there is little reported experience of its use in the particular conditions pertaining in Britain. Its use for the remediation of material of the quantity, physical nature and degree of contamination of the sludges at Grassmoor has, as far as Derbyshire County Council are aware, never been reported previously. Clearly, given the significant costs that would be incurred in the full-scale treatment of these sludges, neither Derbyshire County Council nor the funding agency could commit themselves to this course of action without obtaining some data on the effectiveness of this approach. It was on this basis that the commissioning of trials in advance of commitment to the main works was determined.

2.9 CONTEXT FOR THE TRIALS

From the above it can be seen that the biological treatment trials are directed towards the economic solution of the particular remediation problem relating to the lagoon base sludge material, within the context of a larger integrated reclamation scheme. The decision to carry out biological treatment trials was taken by Derbyshire County Council on the basis that:

i. it was considered likely to provide an economic solution to the particular problems posed by the remediation of this material

ii. a treatment solution was considered likely to minimise the long-term liabilities to the County Council from these materials

iii. the destruction of the toxic contaminants by a biological process was considered to represent the best environmental option, subject to the technical and cost effectiveness of the methods being demonstrated within the trials.

3 Management of the treatment trials

3.1 EVALUATION OF TREATMENT OPTIONS

As described in Section 2.8, several possible treatment technologies were considered. These included biological treatment, extractive technologies, thermal or incineration technologies, and solidification/immobilisation approaches. Extractive technologies such as solvent extraction and soils washing were considered to be unlikely to be effective for treatment of the sludges due to the small proportion of soil particles present, and their fine particle sizes. Furthermore, the significant volume of hydrocarbons within the sludges would still have had to be disposed of by some other means, probably incineration. Similar considerations applied to the thermal desorption approach. Incineration, though likely to be effective, was regarded as undesirable both on grounds of expense and the probable local opposition to the establishment of incineration plant on the site. Transporting large quantities of the material off site was also considered undesirable.

Of the biological treatments considered, *in-situ* methods were regarded as being unlikely to succeed due to the difficulties of introducing oxygen and nutrients to the sludges. There was also concern over the possible inadvertent initiation of combustion in the colliery spoil tips by the introduction of oxygen by air sparging. Of the *ex-situ* approaches, two generic approaches were considered: treatment as an aqueous slurry in a bioreactor, involving the formation of a suspension of the sludge in an aqueous medium, or a soil-based approach, where the sludge would be mixed with soil material, ameliorants and nutrients, and treated within a treatment bed or landfarm, or within a composting pile or biopile. In technical terms, the slurry bioreactor or soil-based approaches appeared to have similar chances of success and it was considered that the process of procurement should allow both approaches to be considered and evaluated at field-trial stage.

3.2 PROJECT MANAGEMENT

The Grassmoor Lagoons reclamation scheme is part of the Derbyshire County Council land reclamation programme, a rolling programme of derelict land reclamation formerly funded under the Derelict Land Grant regime, now administered by English Partnerships (the trading identity adopted by the Urban Regeneration Agency). The overall scheme management and design is carried out by Derbyshire Consulting Engineers, a division of the Environmental Services Department of Derbyshire County Council.

The effective integration of the treatment-based remediation of the lagoon base sludges with the remainder of the reclamation scheme required that the management of the scheme overall was carried out in an integrated fashion by Derbyshire Consulting Engineers. However, treatment-based remediation methods at the technical level require particular specialist knowledge and competencies. Derbyshire County Council therefore determined that, to meet these criteria, the County Council should contract directly with a specialist soil treatment contractor.

The procedure for selecting the treatment contractor and the details of the procurement process for the treatment trials, described below, are summarised in the flowcharts in Appendix A1. The result of the selection process and procurement procedure was that

Derbyshire County Council contracted with the specialist firm Celtic Technologies Ltd. to carry out the treatment trials. Celtic Technologies subcontracted Robertson Laboratories to carry out their chemical analytical work.

Derbyshire County Council also decided that independent verification of the remediation would be desirable and contracted with a chemical testing laboratory, Alcontrol UK Ltd., to carry out independent chemical testing on samples taken by Derbyshire Consulting Engineers during the trials. Finally, Derbyshire Consulting Engineers contracted with CIRIA to carry out the reporting of the trials for CIRIA's case study demonstration programme, and this work package included an element of further chemical testing which was to be carried out under the contract between Derbyshire County Council and Alcontrol UK Ltd. The roles of the various organisations involved in the project are summarised in Box 3.1.

Box 3.1   *Project management roles*

Client	Derbyshire County Council (part of land reclamation programme)
Funding agency	English Partnerships
Scheme designer and project manager	Derbyshire Consulting Engineers (internal Consultancy of Derbyshire County Council)
Contractor for the trials (partial design and implement)	Celtic Technologies Limited
Contractor's chemical testing sub-contractor	Robertson Laboratories
Client's independent chemical testing sub-contractor	Alcontrol UK Limited

3.3   SELECTION PROCESS FOR THE TREATMENT CONTRACTOR

At feasibility stage, an initial screening exercise had been carried out by Derbyshire Consulting Engineers, based on examination of technical publications and marketing literature obtained from companies offering services in the field of soil treatment and bioremediation. This initial screening process was carried out with the aim of determining whether the services required were available within the UK market, and also to ascertain the nature of the projects carried out in the UK by these companies. Following the decision in principle to proceed with the treatment trials, a formal advertisement seeking expressions of interest was placed in October 1996 and the twenty six companies that responded to the advertisement were invited to complete a structured questionnaire upon which initial assessment and pre-qualification could be based. A copy of this questionnaire is enclosed at Appendix A2.

The questionnaire consisted of three sections: the standard application form for Derbyshire County Council select tender lists (focusing mainly on financial stability and health and safety record), the Derbyshire Consulting Engineers' ISO 9000 quality assurance questionnaire and a project-specific questionnaire which sought information concerning the nature of soil treatment services offered by those who responded to the advertisement, their degree of experience, their technical staffing details and their experience with the particular type of materials present at the Grassmoor site.

14 companies responded to the questionnaire. Initial screening was carried out by assessing the questionnaires against the six criteria listed in Figure 3.1.

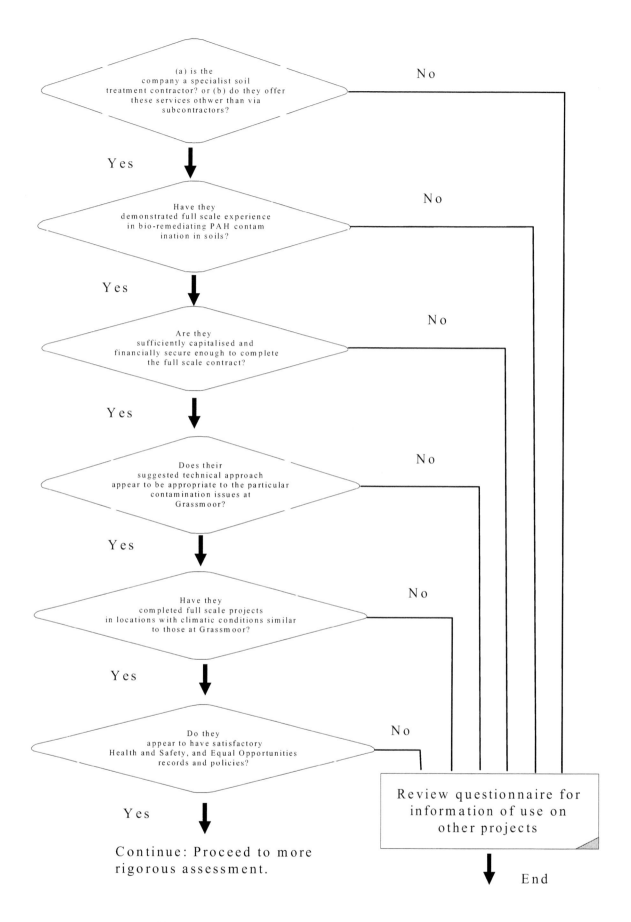

Figure 3.1 *Flowchart for the initial screening of questionnaire responses*

This screening reduced the field to a "longlist" of seven companies, further reduced to the final shortlist of four following detailed consideration of the questionnaire responses from the seven selected companies. This consideration included a financial check carried out by the Treasurer's department of Derbyshire County Council. This check revealed a particular issue concerning the small size and low capitalisation of many companies active in this field. The low capitalisation may impact on the willingness of some clients such as public authorities to award contracts in the case where financial stability criteria are observed. The completion of this process resulted in a short list of four specialist companies or joint ventures for the field trials, two of which were proposing the use of soil-based treatment methods and two of whom were offering aqueous slurry bioreactor methods.

It was a highly structured selection procedure, driven by the necessity of making a selection on technical as well as financial grounds in a field where few companies have a significant track record and where, technically, their approaches differ widely. At the same time it was necessary to follow an equitable and transparent selection process. It involved substantial effort and time in the preparation and despatch of the questionnaires and in assessing the information obtained from the respondents. There was also the total effort expended by the companies responding to the questionnaire.

The quality and detail of the responses to the questionnaires varied considerably. It may be that some of the companies who put little effort into replies and perhaps relied on standard statements of qualification could have been at a disadvantage in the assessment procedure, compared with companies that took greater care to focus their response.

If treatment-based remediation methods are to become more widely adopted, and if public bodies and authorities were to follow a similar procurement procedure, an element of standardisation of technical and financial questionnaires, and of the procedure for obtaining responses to these questionnaires, would be advantageous, both to the selecting organisations and to the organisations responding to the questionnaires. It may even be advantageous for a central public or non-governmental organisation active in the land remediation field to establish and maintain a database of services offered by specialist soil treatment contractors. The database could then be made use of by organisations wishing to procure such treatment-based remediation services.

3.4 PROCUREMENT FOR THE TREATMENT TRIALS

Taken in the context of the reclamation scheme as a whole, the options for procurement of the works varied from the award of a large contract encompassing the whole of the reclamation scheme, incorporating the treatment trials and the treatment works within the larger contract, to the award of a series of smaller contracts independently managed by Derbyshire Consulting Engineers.

Incorporation of the treatment of the lagoon base sludges within a larger contract had several drawbacks. In the absence of results from treatment trials it would have been extremely difficult for a tenderer to determine the time, costs and programming implications of the treatment on the contract as a whole. This would have represented significant commercial risk to the contractor, which would be expected to be reflected in the tender pricing for the contract as a whole, or to increase the likelihood of significant and costly claims for any disruption from the contractor. The specialist treatment sub-contractor would either have had to be nominated or to have been selected by the main contractor rather than by Derbyshire Consulting Engineers.

The separation of the different elements of the work into separate work packages also had drawbacks, in particular the necessity of managing and co-ordinating several smaller work packages. There were also particular difficulties concerning the relationship between the treatment trials and the treatment main works. If the treatment trials were procured independently, and the selection of the trials contractor was based purely on a competitive price for the trials, then Derbyshire County Council could have become "locked" into the particular treatment method or technology supplier used in the trials. At the least, the successful bidder for the trials could gain a potentially significant commercial advantage at the stage of tendering for the main works.

On the other hand, basing the selection on a tender for the main works, with trials being carried out during the early stages of the contract, might result in a contractual commitment to a method that proved unsuccessful at trial stage, and would in any case be difficult and risky for the specialist soil treatment contractor to price in the first instance. Nevertheless, it was considered by Derbyshire County Council that the procurement of the trials, and the assessment of the technical and financial proposals from the short-listed prospective partners, should be based at least in part on their proposals for the main works.

Derbyshire County Council's original intention was that two trials would be carried out: one into a soil-based method and the other into a slurry bioreactor approach. From these "back-to-back" trials, one contractor was to have been selected to carry out the main works, provided that the trials succeeded in demonstrating the economic treatment of the contaminated material to an acceptable residual level of contamination. Box 3.2 indicates the key attributes of the treatment trial procurement strategy.

Box 3.2   *Key attributes of the treatment trial procurement*

- Specialist treatment contractors invited to submit proposals and to tender have been selected on the basis of responses to a public advertisement and analysis of their replies to a questionnaire.
- The contracts for the trials and the main works are partly based on "design and implement" and therefore parts of the contract documentation will be written by the tenderers.
- As a result, while indicative timescales and end product requirements are quoted in these documents, the tenderers must propose their own timescales and end product requirements as part of their submissions.
- Due to the requirement for proposers/tenderers to prepare elements of design as part of the tender at their own risk, the tender shortlist is limited to four tenders.
- The Standard Conditions of Contract on which these documents are based are from the Engineering and Construction Contract (New Engineering Contract, 2nd edition) suite of documents.
- The payment method for the trials contract is a priced activity schedule; the main works will be paid on the basis of a target cost contract with an activity schedule.
- Assessment of the proposals and tenders shall not be based exclusively on the tendered price.
- The intention of Derbyshire County Council was to procure two simultaneous trials of different treatment methods which would be compared on the basis of time, end product achieved and cost, and on the basis of these comparisons to procure a single contract for the main bioremediation works.
- In order to maximise the benefit of the trials, the contract for the main works may be varied by agreement at the completion of the trial stage.

Due to the complex issues raised by this procurement procedure and the unusual and specialised nature of the work, the conventional process of tendering against a fixed design and series of contract documents was not considered appropriate. As indicated in Figure 3.2, documentation for the contracts for both the trials and the main works were therefore based on the Institution of Civil Engineers' Engineering and Construction Contract (New Engineering Contract, 2nd edition) (Institution of Civil Engineers, 1995).

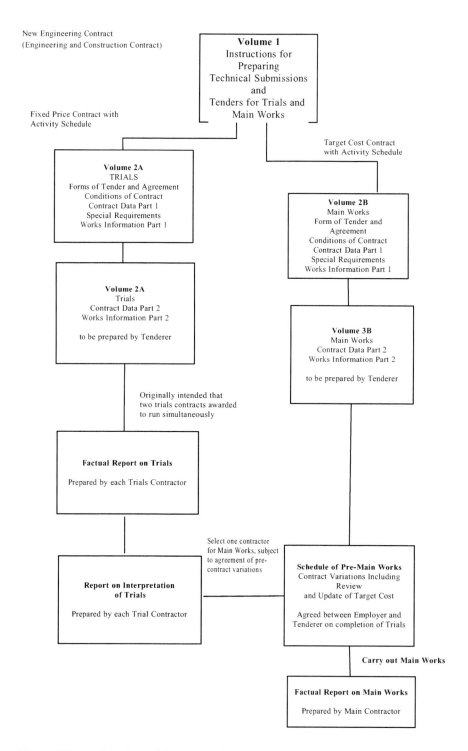

Figure 3.2 *Structure of documents for the treatment trial procurement*

This allowed elements of contractor design to be incorporated, and generally allowed more flexibility in preparation than the traditional civil engineering forms of contract. The trials tender was to be based on a fixed price/activity schedule form of contract, while the main works would be based on a target cost/activity schedule form of contract (Perry *et al*, 1982).

The main works target cost was to be based on the tender for the main works submitted at the same time as the tender for the trials, but subject to adjustment by agreement to take the results of the trials into account. Thus, in contractual terms, the tender for the main works was essentially indicative, although the financial and technical proposals for the main works were taken into account in the selection of the treatment contractor.

The selection of the treatment contractor was based on a weighted assessment matrix of six selection criteria, as indicated in Table 3.1, each criterion had a weighting assigned, against which the quality and economy of proposer/tenderer's submission would be ranked. The total score from each proposer/tenderer could then be assessed as the sum of the products of the weighting and the ranking for each of the six criteria.

Table 3.1 *Assessment matrix for the tenders for the trials*

Selection criteria	Weighting W %	Ranking of proposal R %	Score W × R %
1. Adequacy, sufficiency and detail of design for main works	20		
2. Economy and sufficiency of definition of costs identified in indicative main works tender	20		
3. Design and planning of trials and trial methodology to resolve uncertainties and identify economies in main works	20		
4. Time, cost, achievability of end product and sufficiency of test/ monitoring protocols for trials	20		
5. Time proposed for main works in indicative proposal	10		
6. Area of site proposed for main works in indicative proposal	10		
Sum of columns	**100**		

The tender/proposal with the highest aggregate score will be selected, provided that the minimum ranking percentage for any criterion shall be 50 per cent. Achievement of a lower ranking may lead to the proposal being excluded from further consideration: alternatively the proposer may be asked to clarify the proposal or to provide additional information.

The selection and procurement procedure undertaken in selecting the treatment contractor was a complex and highly structured exercise. Its main benefit was in demonstrating both within the County Council and to the agency (English Partnerships) administering the grant-funding regime that an equitable and transparent competitive procurement procedure was being followed in the selection.

3.5 RESULT OF SELECTION PROCEDURE

Of the four companies or groups shortlisted, three submitted proposals and tenders, two for soil-based approaches and one for a slurry bioreactor approach. It was apparent that the soil-based approach was likely to be more economic and achievable in a shorter timescale and therefore it was decided not to proceed with a slurry bioreactor-based trial. The two soil-based proposals, although having somewhat different approaches, were closely comparable in assessment criteria rankings. Based on these rankings, Celtic Technologies Ltd. was eventually appointed as the contractor for the treatment trials in September 1997.

4 The treatment trials

4.1 PROGRAMME OF WORKS

Celtic Technologies proposed to commence sludge characterisation and initial set-up of bench-scale trials commencing early in September 1997. A period of about two months was allowed for completing and reporting these trials. Meanwhile, preparation for the field trial was to commence late in September 1997, with construction of the treatment bed commencing in the middle of October 1997. Some four weeks into the field trial, the treatment bed was to be split and half of the material from the bed formed into a biopile. Thus, for the later part of the field trial, the treatment bed and biopile variants of the soil-based remediation were to be trialled effectively back-to-back. Final sampling of the treatment bed and the biopile was to take place in early January 1998, with factual and interpretative reporting being presented in mid-February 1998.

A start late in the year of 1997 was necessitated by delays in obtaining necessary approvals to continue with the scheme. Consequently, it was necessary to attempt to shorten the period between commencement of the bench-scale trial and commencement of the field trial, in an attempt to avoid starting the latter in the winter. Treatment bed preparation and mixing in conditions of inclement weather would not have been desirable, although it was considered that once established, the microbial action in the bed and biopile would be self-sustaining. Assessment of the winter performance of the treatment system was considered to be an important facet of the trials.

Adherence to the programme described above was not possible, due to a further delayed start and to the necessity to complete and evaluate bench-scale trials before finalising the treatment bed composition for the main works. As a consequence, it was not possible to commence treatment bed construction until the third week in November 1997. This coincided with the onset of a five-week period of wet weather, which created a series of problems for the trials, as described in more detail in Section 4.5. The difficult weather conditions combined with difficulties over the supply of materials required for treatment bed construction resulted in a total delay in excess of ten weeks to the trials. Consequently, addition of the final increment of sludge did not take place until early February 1998 and the final sampling of the treatment areas could not take place until mid-April 1998. The field trials were thus eventually carried out totally during the winter months.

4.2 FURTHER SLUDGE CHARACTERISATION

Limited sampling of the lagoon base sludges, and some chemical testing, had been carried out at the ground investigation stage. Results from samples taken at the site investigation stage were variable, but indicated concentrations of PAHs (as total PAH stated on a dry solids basis) varying from 0.1 per cent to 20 per cent, mineral oil contents of 2–38 per cent, and the presence of VOCs, cyanides, ammonia and sulphate.

Work on further sludge characterisation was intended to allow further evaluation of the characteristics of the sludge on the site, in particular from lagoons A and A1, where there was expected to be the greatest proportion of lagoon base sludge. The further characterisation (discussed in Section 5.2) confirmed the variability of the sludge, and also revealed particular analytical problems associated with the low solids content and the high content of organic compounds. In particular, the reporting of concentrations of contaminants on a dry solids basis can be misleading when dry solids form only a minor proportion of the sludge. Appendix A3 provides a summary of the sludge characterisation results.

4.3 BENCH TRIALS

The purpose of the bench trials was primarily to establish the optimum blend of sludge to other additives. The bench trials were carried out specifically in relation to the materials and contaminants present on the site and also the available imported ameliorants and additives. The additives were selected primarily in order to create a soil-like structure within the treatment bed material (ie to change the physical nature of the sludge), as well as to minimise volatilisation and offensive odour. They were also selected as fertilising nutrients to improve the available nitrogen and phosphorus content of the treatment bed material and enhance the rate of growth of micro-organisms.

Box 4.1 details the additives selected by Celtic Technologies on the basis of the particular conditions on the site. They should not be regarded as applicable to other sites or contamination scenarios.

Box 4.1 *Additives and their purposes*

	Material requiring treatment was a sludge composite from lagoons A and A1
Organic soil	Low grade peat as soil structure enhancer and to minimise odours and volatilisation
Wood chips	Degradable organic material to assist biological action
Colliery spoil	Bulk material available on site
Crushed limestone	Fine graded material to maintain alkaline conditions.
Inoculum	Soil from lagoons site assumed to contain microbial flora acclimatised to contamination present
Moisture	Maintained at 17.5–20 per cent, based on BS 1377 definition, ie (weight of water x 100%) / (weight of dry solids)
Aeration	By mechanical mixing
Inorganic fertiliser	As nutrients for microbial actions to improve available nitrogen and phosphorous content.

In total, six separate microcosms were prepared, treatments T1–T3 employing different mixes of sludge, additives and nutrients, and treatments T4–T6 investigating the effect of adding sludge in three separate increments (see Table 4.1)

Microcosms were prepared and monitored in a laboratory in the University of Wales, Cardiff, by Celtic Technologies and were incubated and maintained at a temperature of 15–20°C for a period of seven weeks. Chemical analysis was carried out at the start of the bench-scale trial (time = 0 days), after four weeks for a limited analytical suite (time = 29 days), and at the end of the bench-scale trial (time = 42 days). Analytical results of the bench scale trial are described in Section 5.

Table 4.1 *Composition of the bench-scale treatment microcosms*

Treatment number	1	2	3	4	5	6
Sludge (% vol)	25.00	25.00	18.00	7.00	13.00	18.00
Organic soil (% vol)	25.00	–	27.00	31.00	29.00	27.00
Wood chips (% vol)	12.50	12.50	15.00	16.00	14.50	15.00
Colliery spoil (% vol)	37.50	62.50	40.00	46.00	43.50	40.00
Total	**100**	**100**	**100**	**100**	**100**	**100**
Minor additives expressed as % of total:						
Inoculum (% vol)	1.25	1.25	1.40	1.60	1.50	1.40
Crushed limestone (% vol)	1.25	1.25	1.40	1.60	1.50	1.40
Inorganic fertiliser (as total N w/w)	0.50	0.50	0.50	1.00	1.00	1.00

4.4 MECHANIC HANDLING TRIALS

The semi-liquid nature of the sludges presented potential problems in their handling and transport to the site of the treatment bed. These problems had the potential to have a significant cost impact at the main works stage. The toxic nature of the sludges also gave rise to concerns over health and safety of site operatives involved in the excavation and transport process. Attention was thus given to the appropriate methods for handling and transporting the material prior to the commencement of the field trials proper.

Initial consideration had been given to pumping the sludges, which was considered to be possible with appropriate equipment, but this option was not pursued due to concerns over the possible hazards that could be created by pipeline blockages and bursts, and the difficulty of re-starting sludge flow in a pipeline should pump breakdown or downtime occur. The contractor therefore proposed to use conventional excavation and transport methods, which as originally proposed would have involved the excavation of the sludges by small backhoe-loader machine and transport by 1 t dumper trucks. Concern was expressed on health and safety grounds that the semi-liquid sludges might spill from the dumper during transport and come into contact with the dumper driver, or pollute the ground surface. An alternative proposal, to use a tipping high-sided agricultural trailer towed by a tractor, was considered to be more satisfactory from considerations of safety, since the tractor operator would be isolated from any potential spillages.

A mechanical handling trial was carried out on 12 November 1997, following the method described above. This trial showed that the sludges could be excavated using conventional plant, and could be transported satisfactorily in the agricultural trailer, although its semi-liquid nature meant that the trailer could not practically be filled more than half full. One potential health and safety issue was identified during the trial, in that the tailgate of the tipping trailer had to be unlatched by hand, which exposed the operative to the risk of coming into contact with the sludge by spillage prior to tipping of the trailer. The fitment of a hydraulic latch, operated from the cab of the tractor, removed this risk. Videotape recording was made of the operations carried out during the mechanical handling trial, so that it could be an aid at main works stage to operational planning and operator training. Box 4.2 summarises the plant used during the field trial for the three main activities of excavation, transportation and processing.

Despite the semi-liquid nature of the sludge the mechanical handling trial showed that it was capable of standing in a low face during excavation as shown in Figure 4.1. Little drainage of water or free non-aqueous phase hydrocarbons was noted. A method of excavation involving roading in to the lagoons while excavating a retreating face would therefore be possible and is likely to be adopted at the main works stage.

Box 4.2   *Plant used in the field trials*

> **Excavation:** JCB 3CX backhoe loader with "extending" arm.
>
> **Transport of sludge:** Case 125 HP 4 × 4 agricultural tractor towing 7 t tipping trailer with hydraulic tailgate.
>
> **Processing of sludge:** Case agricultural tractor operating a power-harrow

Figure 4.1   *Excavation of sludge in Lagoon A*

4.5   FIELD TRIALS

Celtic Technologies Ltd. originally proposed to establish a 30 m × 30 m treatment bed area on the south tip within the Grassmoor Lagoons site, in the area of the former lagoons D and E. The treatment bed was to be divided into eight sub-areas by the establishment of a control grid of approximately 7.5 m × 15 m. Two of the sub-areas were to have their bases sealed with clay to investigate the impact of restricting under-drainage and to enable leachate to be collected. Sludge was to be added to the prepared treatment bed in three increments, each addition separated by one week. This approach had proved successful in the bench-scale trial, and it was believed that this was due to the micro-organisms having had a chance to become habituated to the high concentrations of organic material being added.

Mixing and aeration of the treatment bed was to be carried out by full depth mixing, using initially a hydraulic tractor-mounted rotovator operated on the treatment bed. The intention was that four weeks from the commencement of the trial, the treatment bed would be divided and approximately 50 per cent of its material would be placed into a biopile to compare the performance of the treatment bed and the biopiling method.

Construction of treatment beds commenced on 17 November 1997 and continued until 25 November 1997. This coincided with a particularly inclement period of weather: around 65 mm of rain fell in the last two weeks of November, approximately 65 per cent greater than the local long-term average for the period. A consequence of the heavy rain waterlogging the treatment bed, and the movement of the tractor on the bed, was that the base of the bed (the infilled lagoon D) became so soft that the tractor became bogged down during bed rotovation. It was therefore not possible at this stage to add further sludge, as this would have decreased still further the workability of the treatment bed. During the first two weeks of December 1997, attempts were made to continue rotovation of the treatment bed to help it dry out to encourage an increase in biological action, but these were inhibited by the continuing wet weather conditions. The layout of the original treatment bed is illustrated in Figure 4.2.

Figure 4.2   *The original treatment bed*

In order to overcome these difficulties, changes were made to the layout of the treatment bed system and the trials methodology. These included the removal of the treatment bed material into a temporary stockpile, the construction of a granular base for the treatment bed area, the addition of further dry wood chips to the treatment bed mixture and the earlier construction of the biopile. These proposals began to be put into effect during the third week in December, with the removal of the treatment bed material to the temporary stockpile and the placing of the granular base layer to the treatment bed.

Further work in re-establishing the treatment bed could not be commenced prior to the Christmas shutdown and so the treatment bed material was left, covered and protected by plastic sheeting, in a temporary stockpile during this period. The part of the original treatment bed which had the clay seal was left in place to monitor the effect of the waterlogging, and did not have the second increment of sludge added.

Work recommenced early in January 1998 with the re-establishment of the treatment beds on the new granular base. It was meanwhile noted that the temperature in the temporary stockpile had increased considerably, indicating that biological action was occurring in the material. However, a delay in delivery of the dry wood chips (for which only one supplier could at that time be located) meant that the completion of the alterations to the treatment bed, and the construction of the biopile, could not commence until the end of January. In the meantime, continued rotovation of the reinstated treatment bed in dry weather had somewhat reduced waterlogging of the material. Following the addition of the wood chips, the biopile was constructed and the rotovation of the reinstated treatment bed was continued on a twice-weekly basis (see Figure 4.3). As a precaution against further waterlogging, the treatment bed was covered with plastic sheeting between rotovation periods.

Figure 4.3   *Rotovation of the treatment bed*

In the third week of February 1998, the second increment of sludge was added to the treatment bed and the biopile. The latter was dismantled and the material spread on the treatment area to enable the further sludge mixing to occur. The biopile was then reconstructed once a reasonable degree of mixing had been achieved. The layout of the biopile is shown in Figure 4.4. Following the second addition of sludge, the beds again appeared very wet and oily. Their appearance steadily altered during the remainder of the trial period, the material in the treatment bed appearing progressively more soil-like in texture and appearance.

Figure 4.4 *Layout of the biopile*

The trial period ended on 14 April 1998, when the final sampling of treatment bed and biopile were carried out. Sampling of the original clay-undersealed treatment bed and of some material remaining from the temporary stockpile from the Christmas break was also carried out to determine if concentrations of contaminants in these materials had continued to reduce. The treatment bed and biopile were left in place, in order that further monitoring and possibly testing could be carried out to determine the longer-term effects on the material from extended remediation.

4.6 ENVIRONMENTAL MONITORING

During the field trials, monitoring for VOCs was undertaken at various locations on the site using a photo-ionisation detector (PID). This monitoring was carried out immediately above the treatment beds, at the boundary of the fenced lagoons site and, latterly, at other locations remote from the country park. The PID measurements indicated generally low concentrations of airborne VOCs. In addition, personal monitoring of site operatives was carried out during the operations of excavation and handling of sludge from the lagoons, and initial mixing into the treatment bed. This personal monitoring consisted of badges containing Draeger tubes, which were analysed for a range of VOCs and showed that occupational exposure levels of site operatives were significantly lower than occupational exposure limits published by the Health and Safety Executive. The environmental monitoring results are summarised at Appendix A4.

4.7 HEALTH AND SAFETY

The scope and scale of the treatment trials was such that the applicability of the Construction (Design and Management) Regulations (Health and Safety Commission, 1995) to the trials was not certain, but it was decided that it was prudent to carry out the field trials within the safety management framework provided by these regulations. A pre-tender Health and Safety Plan, which was of necessity limited in scope since elements of the trials were to be designed by the contractor, was provided to proposer/tenderers at the tender stage. Celtic Technologies Ltd. prepared a construction stage Health and Safety Plan, and all activities on site were preceded by the preparation of detailed method statements. Site operatives were provided with personal protective equipment (PPE), as well as the personal monitoring badges referred to above.

During the early stages of works, involving treatment bed preparation and transport and mixing of sludge, a decontamination unit was provided on site. Cross-contamination of clean areas outside the lagoon area was prevented by the use of a powered jet wash to clean excavation, transport and processing equipment used on the site before it left the site. The lagoons area, in which the trials took place, is fenced off from the surrounding country park, and therefore no temporary site fencing was necessary.

4.8 PUBLIC INFORMATION

Throughout the last few years, considerable effort has been undertaken to prevent public access to the Grassmoor Lagoons site, including the erection and maintenance of security fencing and warning signs. The very act of preventing access to a site which is located within a country park, and the methods used, has consequently advertised to the public that the site is hazardous. The presence of the contaminated and therefore hazardous site has given rise to public concern. This concern has not only been in terms of the location and proximity to housing and amenity areas, but also about when, how and by whom the site will be remediated.

In order to provide information about the wider reclamation scheme proposals and also the bioremediation trials, a public information exercise was initiated. The method of disseminating information about the reclamation scheme and the bioremediation trials took the form of a news sheet (a copy is shown as Appendix A5). The news sheet used a combination of questions and answers about bioremediation along with factual information relating to the overall reclamation scheme. Details of the contractor and project personnel were also included as potential contact points for further information. Copies were posted at various locations around the site and also in local libraries and community centres. General feedback from the public information exercise indicated that the approach used was well received and informative. No adverse comments were received about the information sheet. Responses from users of the Country Park during the trials period were also reported to be positive. Neither the contractor, Derbyshire County Council nor Derbyshire Consulting Engineers received any complaints throughout the duration of the trials. Whether this was as a result of the public information exercise or not, any potential concerns from members of the public about the bioremediation trials were not realised and the public information exercise was considered a success.

5 Results of the trials

5.1 SLUDGE CHARACTERISATION

The characterisation of the sludge in numerical terms by analytical chemical testing was fraught with difficulties. The semi-liquid nature of the material, coupled with its hazardous and noxious properties, made preparation of samples for analysis in the laboratory difficult, and the means by which the laboratories prepared the analytical samples are thus likely to have differed in crucial aspects. As discussed above, the convention of reporting the contaminant concentration as a percentage of the weight of dry solids creates particular difficulties in interpretation of the nature of these sludges, as the proportion of dry solids is comparatively minor (typically 20–40 per cent) and variable. This can create apparently wide variations in the calculated concentration of contaminants.

Further difficulties are created by the hazardous nature of the material in both physical and chemical terms (Sax, 1979), which makes access to the bulk of the sludge impossible and restricts sampling to a narrow and possibly unrepresentative strip along the lagoon edge. During any main bioremediation works contract, the characterisation of the sludge will need to be carried out on a continuous basis, probably at the stage of mixing within the treatment bed or mixing areas.

Table 5.1 indicates the relative methods of defining and quantifying PAH levels in terms of the specific priority pollutant compounds used.

Table 5.2 quotes test results for a selection of key determinants obtained from samples of lagoon sludges. It is notable that the sludges sampled from within lagoon J and lagoon C, which were situated further "downstream" of the original effluent treatment system than were lagoons A and A1, contain significantly lower concentrations of PAHs and phenols than samples from lagoons A and A1. The sludge samples analysed by Alcontrol UK from the latter lagoons, although appearing to have PAH concentrations of about 20 per cent based on dry weight, actually contained 4–7 per cent of PAH as a proportion of the total material by weight (ie the wet weight). Samples analysed by Robertson Laboratories indicated a higher proportion of PAH, averaging 11 per cent, but the 95 per cent confidence limits were of such wide extents that these differences are not necessarily significant, and may merely be indicative of the material variability.

In any case, all figures should be treated with caution because the air-drying process by which the sludges are prepared for analysis may encourage volatilisation of some hydrocarbons and therefore indicate an erroneous proportion of total hydrocarbons. The sludges themselves also appear to exhibit considerable variability in composition. Clearly, it is impractical to determine the composition of the lagoon sludges to within confidence limits closer than a few percent.

Table 5.1 *Definition of EPA 16 and WSR 6 PAHs*

PAH compound	EPA 16	WSR 6
Naphthalene	√	
Acenaphthylene	√	
Acenaphthene	√	
Fluorene	√	
Phenanthrene	√	
Anthracene	√	
Chrysene	√	
Fluoranthene	√	√
Pyrene	√	
Benzo(a)anthracene	√	
Benzo(b)fluoranthene*	√	√
Benzo(k)fluoranthene*	√	√
Benzo(a)pyrene	√	√
Dibenz (a,h)anthracene	√	
Indeno(1,2,3-cd)pyrene	√	√
Benzo(g,h,i)perylene	√	√

* Reported together in analytical results.
PAH concentrations were determined for the 16 "priority pollutant" compounds identified by the United States Environmental Protection Agency (EPA protocol 8100). The subset of six compounds, generally the higher molecular weight compounds suspected of being human carcinogens, analysed in the context of the UK Water Supply Regulations (WSR 6), can be related to allowable leachability values.

Table 5.2 *Characterisation of lagoon sludges*

Date	Sample	Description	Total solids (%)	Mineral oils (mg/kg)	EPA 16 PAH (mg/kg)	Mercury (mg/kg)	Lead (mg/kg)	Arsenic (mg/kg)
		Dry solids basis						
16.09.97	GML/1/01	Sludge lagoon A	20.6	328 000	195 000	18.7	128	24.6
16.09.97	GML/1/02	Sludge lagoon A1	42.8	235 000	170 000	31.4	113	72
25.11.97	GML/2/10	Sludge lagoon J	13.0	4500	165	1.65	19.5	346
25.11.97	GML/2/11	Sludge lagoon C	43.4	4380	220	0.46	17.3	60.5
25.11.97	GML/2/12	Sludge lagoon A	20.1	359 000	263 000	15.7	102	23.7
		Total w/w basis						
16.09.97	GML/1/01	Sludge lagoon A	20.6	67 568	40 170	4	26	5
16.09.97	GML/1/02	Sludge lagoon A1	42.8	100 580	72 760	13	48	31
25.11.97	GML/2/10	Sludge lagoon J	13.0	585	21	0	3	45
25.11.97	GML/2/11	Sludge lagoon C	43.4	1901	95	0	8	26
25.11.97	GML/2/12	Sludge lagoon A	20.1	72 159	52 863	3	21	5

5.2 BENCH TRIAL

Analytical testing for the bench trials was the responsibility of the treatment trial contractor and none was carried out by Derbyshire County Council's independent laboratory. The contractor analysed samples from microcosm treatments T1, T3 and T6, treatment T1 corresponding to a 1:3 sludge:additives ratio, treatment T3 corresponding to a 1:4.5 sludge:additives ratio, and treatment T6 composed of the same ratio as T3 but based on the sludge having been added in three increments.

The total PAH concentrations in the microcosm treatments at the commencement of treatment (the "t = 0" point) was approximately 3 per cent (30 000 mg/kg). During the seven week treatment period, concentrations of PAH in treatment T1 decreased to 12 500 mg/kg, compared with 11 600 mg/kg for treatment T3 and 5600 mg/kg for treatment T6. This confirmed that the staged addition of the sludge appeared to encourage the biological processes because of the change in the physical conditions of the treatment material. Wet conditions with little gas transfer and poor drainage properties inhibit microbial activity significantly.

This staged-addition approach was adopted for the field trials, although in the field trial only two additions of sludge were possible as the moisture condition of the materials in the field was such that further sludge addition would have effectively waterlogged the treatment beds. The degradation progress during the bench trial is indicated in Figure 5.1 in terms of diesel-range hydrocarbons (DRHs) and PAHs.

The bench trials were successful in confirming that the destruction of the high initial concentrations of PAHs by biological action was possible, albeit in the temperature- and climate-controlled environment of a university laboratory. Reductions in the more recalcitrant four and five ring PAH compounds such as pyrene and benzo(a)pyrene of a statistically significant magnitude were also noted, although the rate of degradation of these compounds appeared to be somewhat slower than that of the lighter compounds such as naphthalene. The bench trials demonstrated that, in an appropriate environment, the sludges from Grassmoor Lagoons were capable of being remediated by biological treatment methods.

Analytical chemical test results, provided by treatment contractor Celtic Technologies Ltd., are reported in Appendix A6 to this report.

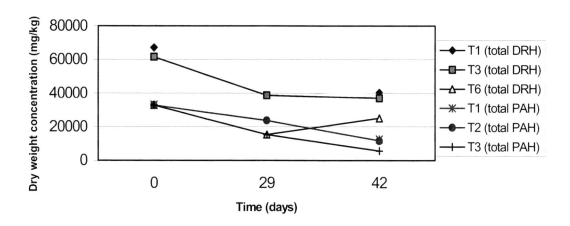

Figure 5.1 *Degradation of DRH and PAH content during the bench trials*

5.3 TREATMENT BED/BIOPILE DURING SLUDGE ADDITION

The field trials were subject to separate sampling and analytical testing by Derbyshire County Council and the treatment contractor. The analytical approaches were different, the contractor concentrating on establishment of a theoretical "t=0" concentration adjusted to take account of the phased sludge addition, while Derbyshire County Council samples were analysed to determine actual concentrations before and after subsequent sludge additions. The results from these different approaches are not inconsistent, but combining them for the purpose of statistical analysis is not practicable. For the purposes of this study, the Derbyshire County Council results based on chemical testing by Alcontrol UK, have been used. Test results and details of the samples taken are given in Appendix A7.

Monitoring of the conditions in the treatment bed and biopile was carried out during the trials by Celtic Technologies Limited. The purpose of the monitoring was to confirm the temperature within the beds, the presence of VOCs and the aerobic nature of the degradation process. Monitoring results are quoted in Appendix A8.

The initial layout of the field trials as shown in Figure 5.2, consisted of a 30 m × 30 m square treatment bed area, subdivided for monitoring and sampling purposes into eight sub-areas, each 7.5 m × 15 m in size. Two of these areas, designated A1 and B1 for sampling purposes, had a clay base installed, the original intention being that this would enable any leachate from this area to be collected and analysed.

The treatment bed was prepared by spreading the colliery spoil, followed by the organic soil (low grade peat) and the wood chips, and rotovating this mixture in an attempt to create as much homogenisation as possible prior to the addition of the sludge. Following this initial rotovation, and prior to the addition of the sludge, a round of samples was taken to determine the baseline chemical characteristics of the treatment bed materials.

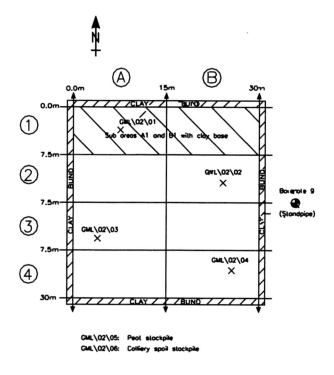

Figure 5.2 *Layout of the original treatment bed*

It is noteworthy that a certain amount of mineral oil and PAH contamination was observed in the treatment bed materials even before the addition of the sludge. This originated from the colliery spoil material, which was obtained from within the Grassmoor Lagoons site and had been in contact with PAH contamination prior to the commencement of the trials (see Table 5.3).

A certain amount of heavy metal contamination was observed, particularly of arsenic, which is not uncommon as a naturally occurring contaminant in coal measures strata in the area. However, the presence of the heavy metal contaminants in both the treatment bed materials and the sludges does create a difficulty in using the concentrations of heavy metals as an index to the dilution effect. The concentration of mercury can be used for this purpose, as mercury concentrations are generally low within the treatment bed materials themselves, but are notably higher within the lagoon base sludges. However, mercury may be more prone to leaching over time than arsenic or lead, depending on the chemical form present.

Table 5.3   *Initial contamination concentrations*

Contaminants	Treatment bed (mg/kg)
PAH (EPA 16)	13
Arsenic (soil)	99
Lead (soil)	88
Mercury (soil)	0.8
Mineral oils (soil)	425

Following addition of the sludge and the initial mixing process, a further round of samples was taken in order to determine the concentration of contaminants after the first addition of sludge. Chemical analysis results at this stage indicated that the treatment bed, as prepared, had mean concentrations of PAH of approximately 8400 mg/kg, and of mineral oils of 5900 mg/kg. Although at this early stage, prior to effective mixing of the treatment bed, the variance on the concentrations was significant, phenols were present in low concentrations (8 mg/kg).

In the difficult weather conditions at the time of the first sludge addition, the treatment bed became waterlogged. Processing of the material became impractical with the softening of the treatment bed subgrade, which caused the tractor to bog down. This necessitated removal of the material from the treatment bed and placing it in a temporary stockpile while the treatment area was reconstructed with a granular base.

In order to reduce the waterlogging of the material, it was decided that a further increment of dry wood chips would be obtained to add to the material when it was re-spread on the treatment bed. During the period that the material was placed in temporary stockpile, it was covered with plastic sheeting to protect it from the weather and, though it remained very wet, a significant build-up of temperature was noted within the stockpile, indicating the presence of biological activity.

Despite the absence of any measures to encourage aeration of this temporary stockpile, such as the porous pipes that would normally be constructed into a biopile, and despite its being covered with plastic sheet, monitoring showed that the temporary stockpile remained aerobic throughout this period.

Following the delayed delivery of the dry wood chips at the end of January, the material from the stockpile was removed and formed into a reconstructed treatment bed approximately 8 m × 16 m in dimension (see Figure 5.3) and a biopile. Rotovation of the treatment bed continued for a further three weeks before the second sludge addition took place. In order to add the second increment of sludge to the biopile, it was dismantled and the material was spread next to the treatment bed. This temporary bed is referred to in this report as the biopile mixing bed to distinguish it from the treatment bed proper.

Immediately before the second sludge addition, the treatment bed and the mixing bed were sampled in order to determine the degree of degradation that had occurred since the initial addition of sludge. The analytical results showed that despite the non-optimum condition which this material had been in for the majority of the time since the first sludge addition, a significant proportion of PAH contaminants had degraded.

The concentration of total EPA 16 PAH in the material decreased from a mean of 8400 mg/kg to a mean of 3750 mg/kg during this period, and mineral oils from 5900 mg/kg to 4350 mg/kg. Concentrations of PAH and other compounds within the treatment bed were considerably more uniform than had previously been the case, a consequence of the effect of the rotovation and mixing of the treatment bed. Phenols were no longer present at detectable levels.

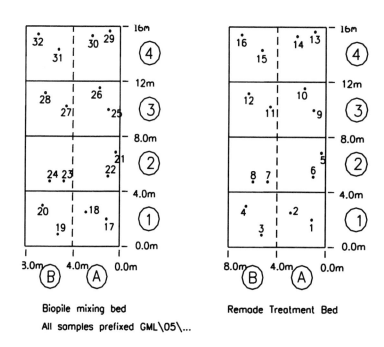

Figure 5.3 *Layout of the revised treatment bed*

5.4 TREATMENT BED/BIOPILE AFTER SLUDGE ADDITION

The second addition of sludge increased the apparent wetness of both treatment bed and mixing bed and it was decided that it would not be practicable to proceed with the third and final addition of sludge within the already extended time scale for the treatment trials. Accordingly, immediately after the initial rotovation and mixing of the second sludge increment, a comprehensive sampling programme on the treatment bed and the mixing pile was carried out in order to be able to achieve a characterisation of the treatment bed prior to the main period of bioremediation so that a statistically valid conclusions could be drawn.

Sampling was carried out on the basis of a technique known as stratified random sampling, two samples being taken from within each of eight sub-divisions within the treatment bed and the biopile mixing bed, a total of thirty-two samples.

At the stage of the second sludge addition, concentrations of EPA 16 PAHs recorded in the treatment bed were on average 11 300 mg/kg, compared with 11 200 mg/kg within the biopile mixing bed. The respective figures for mineral oils were 8800 mg/kg and 10 000 mg/kg, Phenols were generally below 5 mg/kg. Although the concentrations of contaminants appeared to be similar between the material that had been rotovated within the treatment bed and the material that had been contained in the biopile, the visual appearance of the treatment bed suggested that it had a lower moisture content than the material that had been within the biopile. This was consistent with visual inspection of the treatment bed and biopile after the addition of the sludge, when the biopile material again looked seriously waterlogged, compared with the treatment bed material which, while damp and oily, was significantly less wet in appearance.

Toxic metals present (as shown in Table 5.4) in the treatment bed and in the biopile included mercury at a concentration of about 2 mg/kg, lead of about 60 mg/kg, and arsenic at about 90 mg/kg. These values can be compared with those from later analyses to confirm that dilution had not been responsible for the reduction in PAH concentrations achieved.

Table 5.4 *Concentrations after second sludge addition*

Contaminants	Treatment bed (mg/kg)	Biopile (mg/kg)
PAH (EPA 16)	11 324	11 210
Arsenic (soil)	85	99
Lead (soil)	58	65
Mercury (soil)	1.9	2.2
Mineral oils (soil)	8824	10 031

Leach tests were also carried out on samples taken from the treatment bed and mixing bed, based on the Environment Agency's (former National Rivers Authority) Method 301 (NRA, 1994; Lejeune *et al*, 1996). These indicated leachable EPA16 PAH of approximately 5 mg/l and leachable ammoniacal nitrogen of approximately 340 mg/l. COD values of 760 mg/l indicated a high degree of organic concentration in the leachate water.

The treatment bed, although not the biopile, was sampled at an intermediate point approximately four weeks after the second addition of sludge. Analytical test results showed a 45 per cent reduction in concentrations of PAH contamination during this four-week period to about 6300 mg/kg. Rates of degradation of the individual PAH compounds varied considerably, with the bulk of the early degradation due to the breakdown of naphthalene. Mineral oil concentrations were about 6000 mg/kg.

5.5 TREATMENT BED/BIOPILE AT END OF TRIAL PERIOD

The final sampling for the treatment trials proper was carried out on 15 April 1998, seven weeks after the second sludge addition The sampling for the treatment bed followed a similar pattern to that which had taken place immediately after the second sludge addition. For the biopile, however, which was to be left in place beyond the end of the trials, sampling on a simple area basis was not possible and samples were obtained by the use of a pneumatic window sampler. One window sampling position was located in each control square, and two samples were taken from each window sampler position: one from the top 1.0 m of the window sampler hole, and another below that sample depth generally in the 0.8–1.0 m zone.

A further significant reduction in PAH concentration to a mean of 3350 mg/kg had occurred within the treatment bed during the three-week period since the intermediate samples had been taken. Approximately 70 per cent of total PAH present after the addition of the second sludge increment had been degraded. No evidence was observed of any decrease in the rate of degradation of PAH during this period, although an apparent increase in mineral oils concentration was recorded for which no obvious explanation could be advanced. Phenols were generally reduced to below detection limits. It appears that the degradation processes within the treatment bed were continuing at the time of the final sampling, and further reductions in concentrations of contaminants were confidently expected with a continuation of the treatment process. It was expected that this might also be encouraged by general increases in ambient temperature over the summer months. Figure 5.4 indicates the progress of PAH degradation up to the end of the trial period.

Less degradation was observed of the material within the biopile, even though 30 per cent of PAH concentration had been degraded during this period. Mineral oils remained at a high value of 11 100 mg/kg and phenols, while generally reduced, were recorded at amounts less than 3 mg/kg.

The comparatively slow rate of degradation in the biopile is somewhat surprising, given the degradation of material in the temporary stockpile that occurred after the first sludge addition, but is probably explained by the very wet state of the material as it was placed in the biopile.

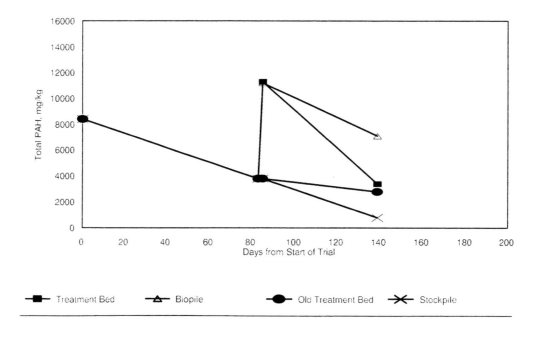

Figure 5.4 *The progress of degradation*

Tables 5.5 and 5.6 provide a summary of contaminant concentrations at the end of the treatment trials. A more comprehensive summary is provided in Appendix A9.

Leach tests results taken on the treatment bed and the biopile indicate degradation of leachable fractions of a similar rate to that of the total concentrations. Of particular note is the continued reduction in ammoniacal nitrogen concentrations recorded in these tests.

Table 5.5 *Concentrations at end of treatment trials*

Contaminants	Treatment bed (mg/kg)	Biopile (mg/kg)
PAH (EPA 16)	3355	7082
Arsenic (soil)	58	82
Lead (soil)	58	60
Mercury (soil)	1.9	2.0
Mineral oils (soil)	7213	11 183

Table 5.6 *Concentrations in the original treatment bed and temporary stockpile at end of trials*

Contaminants	Original treatment bed (mg/kg)	Temporary stockpile (mg/kg)
PAH (EPA 16)	2833	797
Arsenic (soil)	113	92
Lead (soil)	70	65
Mercury (soil)	1.6	1.6
Mineral oils (soil)	5285	3460

As well as sampling the reformed treatment bed and the biopile, the remaining materials of the original treatment bed (the section with the clay underliner) and of the temporary stockpile were also sampled in order to provide an indication of the longer-term degradation of the material which had been subjected to only one addition of sludge. The original treatment bed, probably due to its waterlogged nature and the consequent difficulty of processing it, showed modest degradation to approximately 2800 mg/kg average total PAH. The remainder of the temporary stockpile both showed significant reductions in PAH concentration, having decreased to approximately 800 mg/kg average, and a significant decrease in concentrations of the more recalcitrant pyrene and benzo(a)pyrene compounds. A reduction in leachable PAH concentrations was also noted.

The higher molecular weight PAH compounds were, as expected, more recalcitrant in breaking down, and the reduction in concentration of the four and five ring compounds pyrene and benzo(-a) pyrene from the second sludge addition to the final sampling was around 25 per cent. These compounds, although a comparatively minor proportion of the PAH compounds present in the sludges, are potentially risk driving, however, because they are carcinogenic, a point which has to be recognised in establishing risk-based end points.

Average rates of degradation achieved in the treatment bed following the second addition of sludge ranged from 65 mg/kg/day for naphthalene and 28 mg/kg/day for acenaphthene, to 0.5 mg/kg/day for pyrene and 0.3 mg/kg/day for benzo(a)pyrene. The overall the rate of degradation of total PAH was approximately 146 mg/kg/day, which compared closely with values calculated by Celtic Technologies Ltd. from their theoretical approach. It should be noted that these rates are averaged over a long period of time and take no account of variations in ambient conditions or treatment bed performance.

Celtic Technologies Limited carried out several analyses to determine the nature of the degradative processes, including bacteriological analysis, modelling and monitoring of volatilisation, and calculation of attenuation factors due to dilution. These analyses (see Box 5.1) make clear that non-biological processes can only account for a fraction of the losses in PAH concentration, and also that bacteriological activity and cell counts are high in the material under treatment.

Box 5.1 *Changes to contaminant mass during the field trials*

Aspect of process	Total organic contamination (on a dry weight basis)	
	Concentration reduction (% found in raw sludge)*	Mass reduction (as % of mass of raw sludge)**
Dilution [1]	60–75	0
Biodegradation [2]	15–20	65–80
Volatilisation [3]	0.01–0.1	0.05–0.5
Leaching [4]	0.01–0.05	0.05–0.1

Notes

* Concentration reductions are mainly the result of the relatively high mass of additive required for difficult tars in order to achieve "soil-like structure". Biological treatment is then the same principal cause of reduction in concentration.

** Dilution with additives does not itself reduce contaminant mass. The values assume that there are no losses in the process of dilution. However the process of mixing the additives and sludge to achieve a soil-like structure could have volatilised some of the volatile components.

1. Dilution was used to achieve physical structure.
2. Biodegradation has been assessed based upon mean concentration reductions of a range of contaminants.
3. Air and soil vapour monitoring during the trials has been used to validate soil-air emissions modelling of volatile components.
4. Values are negligible based upon groundwater and leach-test data.

5.6 LONGER-TERM MONITORING

The biological processes leading to the degradation of the PAH compounds were continuing in the treatment bed and biopile at the end of the trial. The results from the old treatment bed and the stockpile indicated that any lower limiting concentration is likely to be considerably less than the concentrations obtained from the treatment bed and biopile in the final testing round.

The treatment bed and the biopile were maintained at the site (Storey, 1998), although treatment bed rotovation was stopped, and limited sampling was carried out at intervals, to determine the effect of a longer treatment period. Appendix A9 provides detailed results of the analysis undertaken after completion of the trials.

The treatment bed was sampled on 12 June 1998, sixteen weeks after the second sludge addition, and nine weeks after the cessation of rotovation at the end of the trials; the results are summarised in Table 5.7. Total EPA 16 PAH results showed a continuing reduction to a mean of 1800 mg/kg, an 85 per cent reduction in PAH concentrations from the levels recorded after the second sludge addition. Mineral oils also showed a fall, reducing to 3500 mg/kg. No reduction was noted in the concentrations of pyrene and benzo(a)-pyrene subsequent to the end of the trials and it is possible that these compounds are less amenable to degradation without the intensive rotovation of the treatment bed.

Table 5.7 *Concentrations in treatment bed sixteen weeks after second sludge addition*

Contaminants	Treatment bed (mg/kg)
PAH (EPA 16)	1816
Mercury (soil)	1.6
Mineral oils (soil)	3494

Further sampling of the treatment bed and biopile was carried out in August 1998, 26 weeks after the second sludge addition and 19 weeks after the cessation of rotovation. A total of 32 samples were taken for analysis; the analysis carried out on these samples was limited in the range of contaminants tested for (see Table 5.8 and Appendix A9). The total EPA 16 PAH results showed a continuing reduction to a mean of 1104 mg/kg for the treatment bed, giving a 90 per cent reduction in PAH concentrations from the levels recorded after the second sludge addition. The total EPA 16 PAH results for the biopile also showed a continuing reduction to 4736 mg/kg.

Table 5.8 *PAH concentrations in treatment bed and biopile 26 weeks after second sludge addition*

Contaminants	Treatment bed (mg/kg)	Biopile (mg/kg)
PAH (EPA 16)	1104	4736

Analysis of results from a further set of 20 samples taken in July 1999 from the treatment bed and biopile are provided in Appendix A9 and summarised in Table 5.9. The total EPA 16 PAH results showed a continuing reduction to a mean of 474 mg/kg for the treatment bed, giving a 96 per cent reduction in PAH concentrations from the levels recorded after the second sludge addition. The total EPA 16 PAH results for the biopile also showed a continuing reduction to 540 mg/kg (95 per cent reduction). Mineral oils reduced to 1859 mg/kg for the treatment bed and 2785 mg/kg for the biopile.

Table 5.9 *Concentrations in the treatment bed and biopile 71 weeks after second sludge addition*

Contaminants	Treatment bed (mg/kg)	Biopile (mg/kg)
PAH (EPA 16)	474	540
Mineral oils (soil)	1859	2785

As can be seen in Figure 5.5, the continued monitoring of the inactive treatment bed following the trials showed that the degradation processes continued in the treatment bed after the intensive rotovation was ended.

The average rate of total PAH degradation over the first nine weeks post trial was 25 mg/kg/day.

Air quality monitoring is to be continued in order to determine a baseline for local air quality prior to the start of the main treatment works. The question of airborne VOCs emerged during the trial as one of some concern to the public in the area, although modelling carried out by Celtic Technologies Limited indicated that only minor amounts of volatilisation are likely. The materials used in the treatment bed appear to have been successful in mitigating odours and VOC volatilisation.

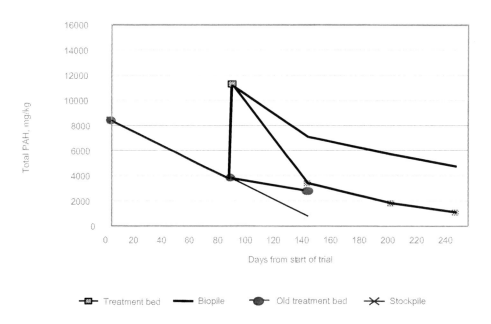

Figure 5.5 *Degradation of PAH continuing after the trials*

5.7 RESIDUAL CONCENTRATIONS RELATED TO EARTHWORKS DISPOSITION AND RISK-BASED END-POINTS

The site is to be reprofiled and areas with significant contamination are to be provided with an infiltration barrier as part of the broader reclamation scheme proposals. The establishment of risk-based end-points will thus be related to the sensitivity of the final disposition of the material (Ferguson and Marsh, 1993; Ferguson and Denner, 1994; Quint and Limage, 1994). Material which is to be left at the ground surface will require a more rigorous end-point, based on the risk of human contact, than material at depth from the surface, where leachability is likely to be more significant, or material placed below infiltration barrier level where regulatory (Environment Agency, 1997 and 1998) rather than risk-based requirements may govern the end-points.

Thus the establishment of risk-based end-points needs to be carried out interactively with the landform design, with the aim of optimising the design of the reclamation as a whole.

6 Cost analysis

6.1 ORIGINAL COST ESTIMATE FOR THE TRIALS

The available options for procurement and the criteria and procedures used to select and employ Celtic Technologies Limited are discussed in Section 3. The criteria used in the selection process identified cost as having a significant influence on the weighting attached to the assessment of the submitted tenders, for both the trials and the main works. Quoted costs and estimates for the main works were purely indicative, primarily because the processes, timescales and treatment targets could only be defined when the results of the treatment trials were known. Therefore the selection criteria, whilst giving due cognisance to estimated costs for the main works, gave a relatively high weighting to the cost of the treatment trials.

The agreed costing schedule for the work undertaken for the trials amounted to approximately £70 000, covering a range of scheduled activities as indicated in Table 6.1.

Table 6.1 *Original costings for the activity schedule*

Activity	Anticipated cost (£)
Set-up	3030
Analysis and evaluation	10 374
Set-up and preparation	7050
Set-up treatment beds	9000
Treatment and monitoring	29 510
Reporting and decommissioning	10 250
Total	**69 214**

In addition to the work undertaken by the treatment contractor, Derbyshire Consulting Engineers appointed Alcontrol Ltd. to carry out independent chemical analysis of the lagoon base sludges and treatment bed/biopile material throughout the duration of the treatment trials. The estimates for this laboratory testing amounted to an additional £15 000.

6.2 COSTS ON COMPLETION OF THE TRIALS

Section 4 of this report discusses the progress of the treatment trials and details the factors that both delayed the progress of the works and, more significantly, resulted in additional costs. The factors that resulted in increased costs are summarised in Table 6.2.

Working within a period of exceptionally high rainfall, with a material already having a high moisture content, resulted in a series of delays and disruptions to the field trials, and consequently led to a general increase in costs. The poor weather conditions led to additional costs in terms of time-related activities because of the slower than anticipated rate of progress in both setting up the site and in completing the trials.

Further increases were incurred for additional plant and materials required for the reconstruction of haul roads, damaged as a direct result of working in poor weather conditions and which was necessary to enable works to proceed.

The actual costs of the field trials undertaken by the contractor were approximately £91 000. The increased costs were primarily as a direct result of the weather conditions, both seasonally poor and with unusually high rainfall levels, affecting the contractor's progress and consequently resulting in delays and/or additional work. Time-related costs amounted to an additional £10 000. A further £9000 was incurred in plant and material costs for the reconstruction of haul roads. Further costs were incurred in purchasing additional dry wood chips and laboratory testing.

Table 6.2 *Events leading to increased costs*

Activity phase	Variance	Implications
Field trials: start-up	Five week period of inclement weather preventing construction of treatment beds	Additional period on site for contractor and equipment. Construction of new haul roads not originally envisaged
	High moisture content of sludge	Dewatering of Lagoon A
Field trials: treatment	Inclement weather	Difficulty in rotovating treatment beds. Problems in lowering the moisture content of the sludge.
	Construction of temporary stockpile	Additional work
	Construction of granular base prior to re-establishing the treatment bed	Additional work
	Difficulties in sourcing dry wood chips	Delays in completing the addition of admixtures

7 Outcomes of the trials

7.1 PROCUREMENT

The procurement process differed from that normally followed by a local authority working within the rules established by a central government funding agency. However, a necessary level of transparency and accountability was maintained albeit at the expense of becoming cumbersome on occasions. It is clear that the necessity of involving the specialist contractor at the stage of scoping the works, and of their participation in the early bench-scale and field trials, raised particular competitive issues.

The approach adopted in this case was initially to link the trials stage to the main works within a strategic framework. This approach was linked to the original proposal to carry out two separate and competitive trials, with only one of the contractors selected for the main works. This concept of selecting the main works contractor on the basis of the results of competing back-to-back trials was not achievable and the issue of competition could therefore not be resolved in this way. Furthermore there remain some residual financial and technical risks which make the continuation to main works within the original framework problematic.

Derbyshire County Council are therefore considering an alternative procurement route for the main works by means of a construction management approach, with the contractor acting as construction manager for a series of smaller and shorter-term contracts encompassing supply of treatment additives, land treatment area preparation and treatment area processing. This approach will reduce the commercial risk to the various parties by avoiding over-commitment of resources at the early stages and will also enable the project participants to share in benefits from experience curve effects as the project progresses.

7.2 TREATMENT METHOD

The technical objective of establishing the optimum combination of treatment method, cost, time and area has been partially achieved. The staged addition of sludge has been shown to be an advantageous means of dealing with initial high concentrations of contaminants, and the use of soil micro-organisms indigenous and habituated to the site was successful. Both rotovated land treatment bed and biopile have shown reductions in contaminant concentration, although in the latter stages the treatment bed appeared to achieve a higher rate of degradation. It appeared from the visual appearance of the soil matrices in the treatment bed and biopile that the key difference was in the moisture condition of the soil. The performance of the biopile at the earlier stage, and the reduction in contamination in the stockpile remainder, suggests that provided the moisture condition of the soil is suitable, a similar rate of degradation can be achieved in the biopile.

Biopiling has the advantage of requiring less processing, therefore having a lower operation cost, but initial mixing and conditioning is best carried out in a treatment bed system.

The final approach adopted for the main works at Grassmoor is therefore likely to be a hybrid of initial more intensive treatment bed processing followed by biopiling. The land treatment areas identified at the Grassmoor site will enable the sludge volumes present in Lagoons A and A1 to be treated in an approximately two-year period. The uncertainty which remains concerning the volume of sludge in Lagoon C requiring treatment (the water in this lagoon is still being drawn down) means that the treatment period overall cannot be finally determined. A further uncertainty concerns whether the sludge-to-solids ratio achieved in the bench-scale trials, which was not achieved in the field trials, can be achieved in the main works stage. The limiting factor in this context is clearly related to the moisture condition of the soil matrix material in the field, and the variability of the sludge.

The success in initiating biological activity despite the unfavourable ambient conditions indicates that the nutrient input requirements were successfully defined. The importance of the temperature condition of the treatment systems is clear. It is apparent from the trials that once biological activity reaches a certain level, the higher temperatures which optimise biological activity are self sustaining. The particular difficulty encountered during the field trials was how to "kick start" the biological activity in order to generate these higher temperatures. This would clearly have been easier in the warmer and generally drier summer months. During winter conditions, the rotovation of the treatment beds creates a cooling effect from the introduction of low ambient temperature air. A conclusion is therefore drawn that the optimum performance from the treatment system may be achieved by operating the treatment bed stages during the summer months and then forming biopiles, which can be left to work over the winter.

7.3 SLUDGE CHARACTERISATION

The exercises in sludge characterisation revealed significant variance in the composition of the sludge, even within the comparatively small area of Lagoon A from which the sludge to be treated was excavated. This creates a degree of uncertainty over allowable ratios of sludge-to-soil matrix. Total volumes of sludge requiring treatment are defined within a wide range, uncertainties remaining over the volume of sludge in Lagoon C. The information gained from the trials has therefore reduced, but not completely eliminated, the commercial and technical risk of the project.

7.4 ENVIRONMENTAL MONITORING

The monitoring of the site during the field trial period and the modelling carried out by the treatment contractor both indicated that the atmospheric impact from contaminant volatilisation was minimal. Other than immediately adjacent to the treatment beds themselves, the measurements of volatile organic compounds were similar to background levels measured outside the site, and generally lower than in locations impacted significantly by road traffic. Monitoring of waters on the lagoons site failed to identify any adverse impacts created by contaminants leached from the treatment bed, although it must be noted that the waters on the site were already significantly impacted by similar contaminants.

The wider reclamation scheme involves the draw-down and management of the site surface water and "perched" groundwater and any leachate from the treatment beds will be controlled within this system.

7.5 FIELD TRIALS

Despite being carried out in difficult conditions, the bioremediation trials at Grassmoor did demonstrate that the difficult sludge material present in large quantities at the Grassmoor site was amenable to remediation by *ex-situ* biological means in practical, field conditions. The field trials have met the six objectives set down in Section 1.2, and have enabled Derbyshire County Council to continue the project to its further phases.

The field trials were commenced during a period of high rainfall and falling temperatures and were progressed throughout the coldest months of the year, but despite these conditions significant reductions in concentrations of contaminants of concern have been measured. During the period of the trials approximately 80 per cent of the total PAH concentration added to the treatment system in the sludge was degraded, at a rate of about 150 mg/kg/day. Further degradation in the treatment bed and biopile material is continuing after completion of the field trials. Carrying the trials out through the winter was not a deliberate decision, it having arisen through earlier programme delays, but it has been of benefit in demonstrating that this form of bioremediation can be carried out under difficult climatic conditions. It may be expected that conditions for biological activity will be more suitable in the warmer and drier ambient conditions of the summer months. The significant reduction in the concentrations of the PAH compounds, achieved despite the climatic conditions, indicates that the sludges are amenable to biological treatment within a soil matrix-based treatment system.

Reduction in concentration of contaminants due to the biological activity in the treatment matrix material was continuing at the time of the end of the trial period. Although the end point concentrations achieved from the treatment beds were not as low as the original limits proposed for the trials, further reductions are therefore likely in time. Monitoring of the treatment system will be continued to determine the rate and magnitude of these reductions. Consideration can now be given to the balance between the sensitivity of the material's final disposition in the regraded landform of the reclamation scheme, the material's risk-based end-points, and the economics and practicality of the biological treatment.

The trials have enabled practical methods of preparing and processing the treatment areas and materials to be determined, and allowed experience to be gained in the practical operation and management of the land treatment system. It is apparent from the trials that the physical nature of the treatment matrix, particularly relating to moisture content and physical structure, is critical to the success of the approach. In this context, the staged addition of the sludge is an important benefit because as well as allowing the micro-organisms to habituate to the high concentrations of contaminants in the treatment system, it allows the suitable physical condition of the treatment bed to be maintained and managed. System optimisation at main works stage may allow several successive increments of sludge to be added to the treatment beds, creating economies in the import of treatment bed construction materials.

The independent verification and audit testing of the process was of benefit at the trial stage as it enabled the treatment contractor to concentrate on the analysis required for control and management of the treatment system, while the independent testing could be focused on auditing the performance of the system. Clearly there is a cost penalty to this duplication of analytical effort, and it is likely that during the main works the use of an independent laboratory would be limited to occasional verification check testing.

8 Conclusions from the case study

8.1 MANAGEMENT PROCESS

The sludge material present at Grassmoor is typical of the coal tar waste streams which arise from coal carbonisation, and is similar in physical nature to the heavier waste streams from crude oil refining. These materials present particular disposal or treatment problems because of their semi-liquid nature, which means that the conventional means of dealing with aqueous or solid wastes are not appropriate. Although traditionally co-disposal in landfill has been used for similar materials, this is limited to small quantities. This option is unlikely to be available under forthcoming EU legislation. The alternative of high temperature incineration is extremely expensive and can be problematic due to poor public perception and the air pollution risk. Thus, a biological treatment approach for such materials is considered as the most appropriate.

However, the nature of the sludges makes it difficult to introduce oxygen, and their structure generally is not amenable to encouraging biological activity. To overcome this problem, alternative biological treatment approaches can be considered: mixing the sludges with water to form an aqueous suspension in which water-based biological activity can take place, or mixing them with solid additives to create a soil-like structure in which soil microbes can flourish. The original intentions of this case study were to trial both methods back-to-back, but the only proposal received for the slurry-based method would have been more expensive and taken longer than the soil-based approach, which was therefore selected for the trials.

While the technical capabilities of this form of *ex-situ* treatment are becoming better understood, the issues of its procurement and integration into the general activities of brownfield remediation are more problematic. The specialist biological treatment activity requires the integration of technical design expertise with ability to organise site processing work – part design, part contracting – which means that it does not fit easily into the construction supply chain. Initial selection of potential treatment specialists is difficult; specialist companies in this field are small, have limited capital bases, track records are generally short and approaches vary significantly between companies. Rigorous like-for-like comparison at pre-tender stage is therefore difficult and time consuming.

The nearest equivalent is the specialist geotechnical ground treatment processes that are commonly carried out by specialist subcontractors. The significant difference is in the difficulty of defining timescales for biological treatment, which makes integration into a conventional construction project more difficult. Inability to define timescales as part of a construction programme in development-led projects translates into commercial risks for clients and contractors, though this is less of a problem at the Grassmoor site which is being reclaimed for amenity end-use.

8.2 RECOMMENDATIONS FOR FUTURE WORK

The approach of carrying out laboratory bench-scale trials followed by field trials before the main works, while reducing uncertainties for the main works, is in itself time consuming. This can be a negative factor in the case of development-led remediation, where compressed timescales are frequently required for financial reasons.

A compression of the preparatory timescale could be achieved by specifying the bench-scale trials as part of the site investigation stage, where desk study indicates that contaminants potentially amenable to biological treatment may be present. In this regard a standardised procedure for the bench-scale tests, which are not in themselves unduly complicated, could be developed. It is likely that as the body of experience of full-scale biological treatment works is developed, it will be possible to omit the field trials stage and define the operational parameters for full-scale work on the basis of the bench-scale trials.

Nevertheless, there are pointers for improved practice and validation that stem from these trials:

- to use controls with bench trials, when there is the possibility of volatilisation of the constituents of the contamination, eg naphthalene
- to consider how to deal with the imprecision of the sampling and analytical procedures (particularly a problem with PAHs in soil)
- to examine the possibility of optimising degradation through further laboratory studies on the effectiveness of nutrient additions, use of surfactants, etc.

The field trial programme described has been carried out on the basis of a practical engineering approach to the particular contamination problem at the Grassmoor site. The focused has been on achieving workable solutions to these problems on site rather than research. While a considerable amount of academic research has been carried out on biological degradation of various contaminants, particularly in the United States of America, much of this research has been carried out in the controlled conditions of designed experiments. There is a need for research to be continued in the practical use of full-scale treatment works to complement and contextualise the practical engineering experience gained during the works.

Such research might include determination of the nature and the dynamics of the biochemical reactions occurring. This should cover: the definition of the mass balance and contaminant fate of the organic materials introduced to the treatment system; the isolation and classification of the microbial flora; and the determination of their interactions and symbioses with each other. This research would be well beyond the scope of normal engineering monitoring and would not be realisable within the budget of any single reclamation scheme. Additional research funding would therefore be required for such activities, though the engineering work/facilities could be used where possible to minimise the costs of this research.

The issue of dissemination of results from practical biological treatment exercises is important to consider. Much of the work carried out into *ex-situ* biological treatment of difficult materials has been in the context of the energy or chemicals industries and a tendency is noted for companies in these sectors, either for competitive or public relations reasons, to avoid any form of publicity of their experiences. This will serve to retard the development and adoption of these methods, which is not to the long-term benefit of such companies that may have significant difficulties with treatment of their current or historic waste streams.

It is suggested that the publication in the technical journals of their experiences would be of long-term benefit not only to the specialist treatment technology providers but also to their particular industrial sectors.

8.3 TRIALS PROGRAMME

The field trial programme has succeeded in its overall aims of demonstrating the effectiveness of biological treatment of the coal tar sludges found at Grassmoor and providing sufficient information to enable full-scale remediation works to be commenced. The combination of high initial concentration of contaminants, difficult physical conditions and sub-optimal climatic conditions present during the trials has not been reported elsewhere in the published literature. Despite these physical constraints, the field trials programme has demonstrated that biological treatment is a more robust approach in these regards than is sometimes assumed. The trials have demonstrated that biological treatment is a viable approach for difficult sludge materials, and system optimisation during the main works is likely to allow further refinement of control parameters and development of the body of knowledge on the approach.

The final and perhaps most significant conclusion drawn from this case study is that it is the difficulty of integrating this form of biological treatment into the brownfield reclamation process, rather than the technical capabilities of the approach, which currently form the greatest barrier to its wider implementation. However, where this issue can be successfully resolved, as may often be possible by planning the project to take the bioremediation activities away from the "critical path", the use of soil-based biological treatment can offer a technically effective and economic solution for treating these difficult waste streams.

A1 Management process flowchart for sludge treatment, as originally proposed

Figure A1.1 *Management process flowchart for sludge treatment, as originally proposed* (on following three pages)

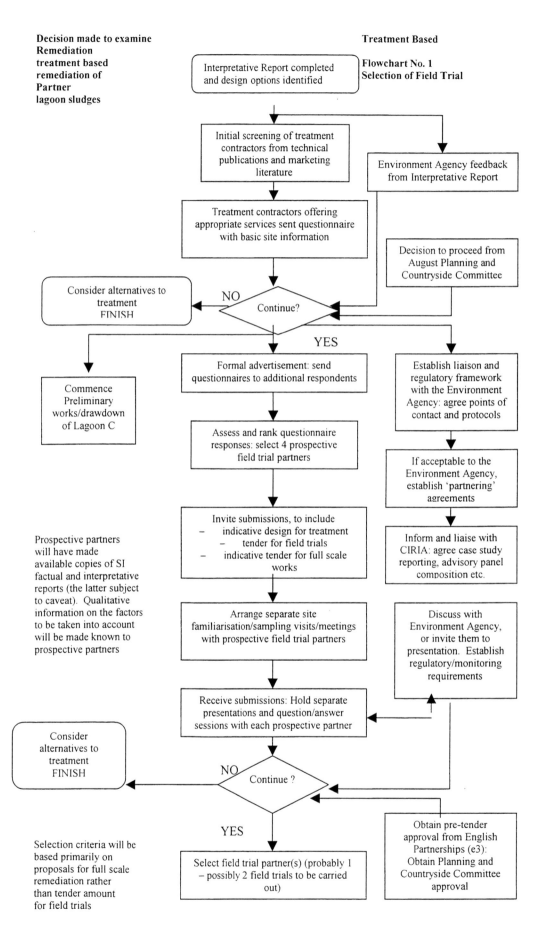

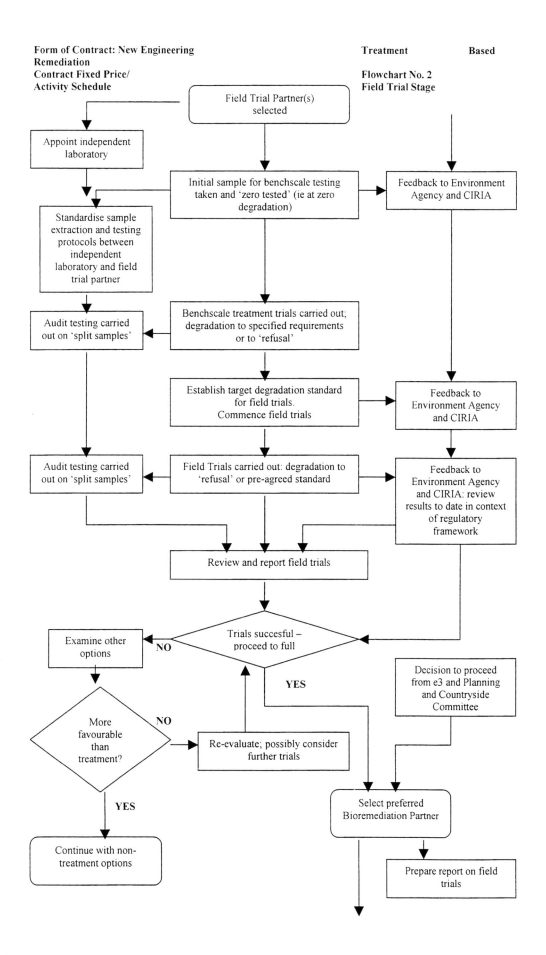

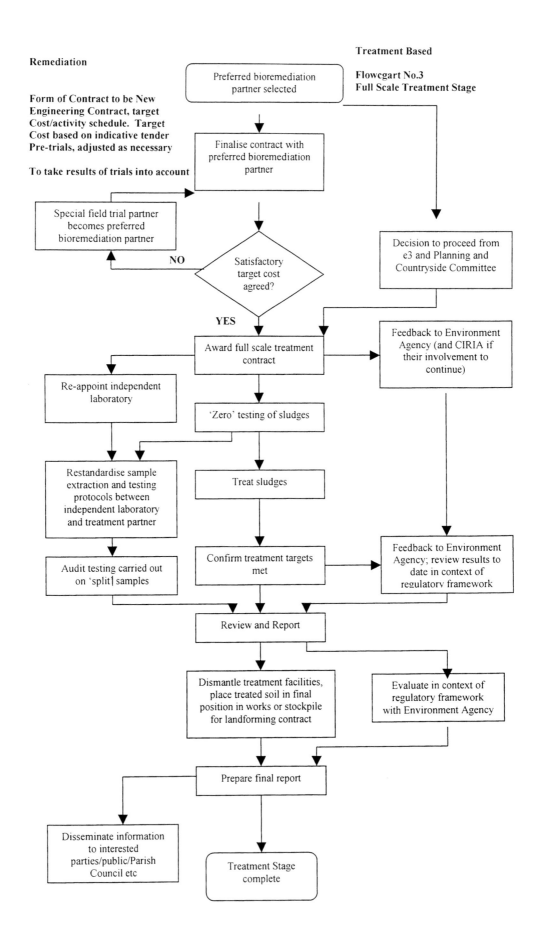

A2   Selection questionnaire

DERBYSHIRE COUNTY COUNCIL CONTAMINATED LAND TREATMENT TECHNOLOGY

Questionnaire

Prepared by: Derbyshire Consulting Engineers, July 1996

Name of Company:

David Harvey
Director of Environmental Services
Derbyshire County Council
County Hall
Matlock
Derbyshire
DE4 3AG

1 Information: responding to the questionnaire: scope of proposed project

Consideration is being given to the use of on site treatment, possibly including bioremediation amongst other options, for dealing with approximately 60 000 m^3 of contaminated soil and sludge at the Grassmoor Lagoons site near Chesterfield, North East Derbyshire. The primary contaminants present in these soils are coal tars and coal tar oils, containing high levels of polyaromatic hydrocarbons and mineral oils. Some indicative analysis results from Lagoon A1, which appears to be the worst contaminated of the lagoons, and other lagoons, are enclosed (see attachment) for information.

The purpose of the questionnaire is to ascertain the capabilities and track records, both financial and technical, of selected companies offering commercial contaminated land treatment and bioremediation services. The questionnaire will also act as a vehicle to enable companies to express their interest in participating in the treatment of these materials. The information supplied in this questionnaire will be considered by Derbyshire County Council in deciding which companies may be invited to tender for future work on this project. **This questionnaire is not an offer of work nor an invitation to tender** and respondents should note that Derbyshire County Council may decide not to employ any of the companies circulated with this questionnaire nor to proceed with the use of on site treatment or bioremediation at this site.

Derbyshire County Council are exploring the possibility of inviting one or more companies to participate in carrying out pilot scale trials, based on the principles of limiting its downside risk in the event of the method failing to achieve the specified targets, and achieving a cost per unit volume comparable with that for more conventional engineering based remediation options, prior to making a decision on whether and how to proceed with full scale remediation at the site.

Two other questionnaires are attached to this questionnaire, dealing with company, financial, health and safety etc and quality management issues. Respondents should note that whilst the standard Derbyshire County Council form *Application for admission to lists of tenderers* is attached and should be completed, Derbyshire County Council does not maintain an approved list of tenderers for treatment based contaminated land remediation or for bioremediation as individual classes of work.

Please be as specific as possible in answering the questions and where appropriate attach supporting documentation. Where insufficient space is provided, or if more convenient for text processing, please answer on a cross-referenced separate sheet of paper.

All three questionnaires should be returned together, **no later than 30th July 1996**, to:

Mr. M. J. Taylor, Principal Engineer - Land Reclamation
Derbyshire Consulting Engineers
County Hall
MATLOCK
Derbyshire DE4 3AG

2 Your company

(Further information under this heading is required in the other two questionnaires)

(a) Please indicate the address(es), telephone, fax and e-mail numbers, of your office(s) dealing with this questionnaire.

(b) Please tell us about your key people: their roles, backgrounds, qualifications and achievements.

(c) Which of the descriptions given below best describes the **major** activity of your company? Please tick one box only, providing clarification as necessary in the space provided.

Consultant/designer	☐
Turnkey process designer-contractor	☐
General civil engineering contractor	☐
Specialist land reclamation contractor	☐
Specialist contaminated land treatment contractor	☐
Specialist bioremediation contractor	☐
Supplier of biological media	☐
Other/Clarification...	

(d) Please indicate if your company is involved in **any** of the activities described below. (Tick as many boxes as necessary)

Consultancy/Design	☐
Turnkey process design-construction	☐
General civil engineering construction	☐
Main contractor on schemes	☐
Sub contractor on schemes	☐
Land reclamation - engineering based	☐
Land reclamation - treatment based	☐
Specialist contaminated land treatment	☐
Specialist bioremediation	☐
Supply of biological media	☐
Other/Clarification...	

(e) Have you acted as subcontractor to a main contractor on any scheme? If so, who was the main contractor?

(f) Which subcontractors have you employed on any of your schemes? What was the scope of their work?

(g) In what areas of (i) the UK, (ii) Europe, (iii) the rest of the world, has your company been involved in carrying out treatment works? Would you consider your company to have any particular geographical centre of gravity in the UK, and if so where?

(h) What do you consider to be your company's particular competitive advantages in the field of contaminated land remediation? What makes you stand out from the competition?

3   Technical: general contaminated land treatment

(a) In what areas of treatment technology for contaminated soils and water, and for how long, has your company operated?

(b) Do you maintain any links for scientific advice, analysis or research purposes, with any independent company, research body or educational institution? Please provide details.

(c) Have any professional papers been presented at conferences or in journals concerning projects and treatment methods that your company has been involved in? Please provide references.

(d) Are you developing or investigating new treatment technologies? Please describe your approach, and how the development is progressing. (We do not expect you to reveal commercially sensitive information.)

4   Technical: specific details

(concerning contaminated land treatment methods potentially relevant to the Grassmoor Lagoons site)

(a) Might any of your techniques be capable of treating soils with the chemical analyses as attached? Please give brief details of these techniques, and list any caveats.

(b) What order of magnitude of residual concentrations might be achieved after (a) 1 month, (b) 3 months, (c) 1 year, (d) ultimately? Please list any caveats.

(c) What indications can you give concerning orders of magnitude/range of costs? Please list any caveats, and indicate any significant factors likely to affect costs.

(d) What secondary contaminated materials, eg solvent, effluent, waste, might be produced and which would require off site treatment or disposal?

(e) What possible negative environmental effects, for example production of gases, leachates, VOCs, odours, etc, might be consequent on your techniques? What steps might you take to ensure their mitigation?

(f) Has your company ever been prosecuted by a Statutory Body (eg NRA) for any pollution or other offence arising from activities in reclaiming contaminated land? If so, please provide details.

(g) Please give examples of projects where you have used similar techniques to that described in this section, on similarly contaminated materials if applicable, including names of clients, dates, tender and out-turn costs, volumes of material treated, residual contaminant concentrations, relationship to other contractors involved in the projects and other relevant information.

(h) Was the effectiveness of treatment on these projects assessed by audit testing or by other methods? Which laboratories were used?

(i) Please provide names, addresses of client companies, consultants, main contractors or public authorities who may be approached for references concerning your work on these projects. (*Note: If any of your clients are prepared to provide an informal reference, but wish to retain confidentiality, we can approach them in confidence: in this case please inform us via separate communication*)

(j) What suggestions would you make for the contractual arrangements and payment mechanism for laboratory bench scale and field pilot scale treatability studies? Would you consider a "no win-no pay" arrangement in the event that previously agreed remediation levels could not be achieved in the field trials?

(k) What would be your preferred timescales for (i) bench scale trials and (b) field pilot scale trials?

5 Contractual

(a) Please provide details/examples of previous contractual arrangements entered into, including names of clients, main contractors, Engineer or Architect and Form of Contract and method of reimbursement.

(b) Please provide details of tender and out-turn costs, reasons for cost over-runs, details of any contractual claims and matters in dispute.

(c) Have you ever been involved in arbitration under a Contract or litigation in connection with a Contract? Please provide brief details.

(d) Please provide names, addresses of client companies, consultants, main contractors or public authorities who may be approached for references concerning your work on these projects. (*Note: If any of your clients are prepared to provide an informal reference, but wish to retain confidentiality, we can approach them in confidence: in this case please inform us via separate communication*)

(e) Do you have any preference for any particular form of contract or contractual arrangement? Please describe any such preferences.

6 Further information

(a) Please provide any further information other than that specifically requested in these questionnaires that you consider may be of use to Derbyshire County Council in reviewing your company and its services.

Attached:
DCC Questionnaire: (to be returned with this questionnaire);
DCE QA Questionnaire: (to be returned with this questionnaire);
Indicative information on the Grassmoor Lagoons site.

A3 Summary of the sludge characterisation results

Table A3.1 *Chemical contaminant testing summary – sludge characterisation*

Date	Sample number	Description	Total solids %	Mineral oils mg/kg	Monohyd. phenols mg/kg	Naphthalene mg/kg
		Dry solids basis:				
16.09.97	GML/1/01	Sludge Lagoon A	20.6	328 000	134	46 700
16.09.97	GML/1/02	Sludge Lagoon A1	42.8	235 000	4050	26 700
25.11.97	GML/2/10	Sludge Lagoon J	13.0	4500	94.2	40
25.11.97	GML/2/11	Sludge Lagoon C	43.4	4380	11.1	120
25.11.97	GML/2/12	Sludge Lagoon A	20.1	359 000	437	80 000

Date	Sample number	Description	Acenaph-thene mg/kg	Pyrene mg/kg	Benzo-a-pyrene mg/kg	EPA 16 PAH mg/kg
		Dry solids basis:				
16.09.97	GML/1/01	Sludge Lagoon A	93 800	1890	586	195 000
16.09.97	GML/1/02	Sludge Lagoon A1	66 300	3120	1390	170 000
25.11.97	GML/2/10	Sludge Lagoon J	54	10	< 4	165
25.11.97	GML/2/11	Sludge Lagoon C	81	5.5	1.8	220
25.11.97	GML/2/12	Sludge Lagoon A	101 000	2800	440	263 000

Date	Sample number	Description	Mercury mg/kg	Lead mg/kg	Arsenic mg/kg
		Dry solids basis:			
16.09.97	GML/1/01	Sludge Lagoon A	18.7	128	24.6
16.09.97	GML/1/02	Sludge Lagoon A1	31.4	113	72
25.11.97	GML/2/10	Sludge Lagoon J	1.65	19.5	346
25.11.97	GML/2/11	Sludge Lagoon C	0.46	17.3	60.5
25.11.97	GML/2/12	Sludge Lagoon A	15.7	102	23.7

A4 Summary of environmental monitoring results

Table A4.1 *Results of personal monitoring for VOCs (in ppm)*

Substance	Long-term exposure limits (8 hour TWA* reference period)	Short-term exposure limits (15 minute reference period)	Maximum concentrations recorded on site**
Benzene	5	–	<0.09
Toluene	50	150	<0.08
Ethyl-benzene	100	125	1.2
o Xylene	100	150	0.1
m/p Xylene	100	150	0.3

Notes:
* TWA Time-weighted average
** Recorded using Draeger ORSA 5 sampling tubes

Table A4.2 *VOC monitoring at site boundaries (continued overleaf)*

Date	Total VOC concentration (ppm) – AM				Air temp (°C)
	Location 1	Location 2	Location 3	Location 4	
11.11.97	0.6	0.8	0.7	0.7	11
12.11.97	0.5	0.6	0.5	0.8	13
13.11.97	0.6	0.4	0.7	0.7	12
14.11.97	0.5	0.4	0.4	0.4	10
17.11.97	0.4	0.4	0.3	0.4	8
18.11.97	–	–	–	–	7
19.11.97	–	–	–	–	9
20.11.97	0.7	0.5	0.5	0.8	10
21.11.97	0.5	0.3	0.2	0.2	10
25.11.97	0.2	0.3	0.1	0.1	9
26.11.97	0.9	0.8	0.7	0.6	8
27.11.97	0.2	0.2	0.2	0.2	9
2.12.97	0.3	0.4	0.3	0.2	8
4.12.97	0.5	0.6	0.5	0.2	3
9.12.97	0.7	0.5	0.5	0.4	8
12.12.97	0.5	0.2	0.2	0.2	8
15.12.97	0.5	0.2	0.2	0.2	6
16.12.97	0.3	0.4	0.3	0.3	6
18.12.97	0.4	0.4	0.3	0.4	2

Date	Total VOC concentration (ppm) – AM				Air temp (°C)
	Location 1	Location 2	Location 3	Location 4	
5.1.98	0.2	0.4	0.2	0.3	5
6.1.98	0.5	0.6	0.7	0.6	8
7.1.98	–	–	–	–	3
9.1.98	–	–	–	–	5
13.1.98	0.5	0.6	0.6	0.8	8
15.1.98	1.0	1.1	0.5	0.9	5
19.1.98	0.4	0.5	0.4	0.4	6
20.1.98	1.1	1.0	1.2	1.0	3
26.1.98	1.2	1.3	1.3	1.5	1.1
28.1.98	1.5	1.3	1.7	1.6	3.1
29.1.98	0.9	1.1	1.0	1.4	-1.3
30.1.98	0.7	1.1	1.5	1.4	6
6.2.98	1.0	1.2	1.4	0.9	9
9.2.98	1.1	1.1	1.0	1.5	9
13.2.98	1.1	1.0	1.5	1.6	19
16.2.98	1.1	1.3	1.4	0.8	9
18.2.98	0.8	0.2	0.2	0.3	11
25.2.98	0.9	0.4	0.4	0.7	11
27.2.98	0.9	0.5	0.5	0.7	10
4.3.98	0.1	0.1	0.1	0.2	7
6.3.98	1.0	0.1	0.1	0.2	4
13.3.98	1.1	0.7	0.6	1.0	9
18.3.98	1.5	0.2	0.2	0.3	10
27.3.98	1.0	0.5	0.5	0.5	13
3.4.98	0.9	0.3	0.2	0.3	9
9.4.98	0.8	0.3	0.5	0.5	5
14.4.98	1.0	0.2	0.3	0.4	7

Note: Maximum = 1.7 ppm, minimum = 0.1 ppm, average = 0.6 ppm

A5   Information sheet

Derbyshire County Council

Grassmoor Lagoons Reclamation Information   November 1997

Water in Lagoons Treated

During the last ten months nearly 50,000 cubic metres (eleven million gallons) of water from the lagoons has been drawn down and discharged to the main sewer. It is treated at Whittington Moor sewage works by Yorkshire Water.

Treatment Trials on Sludges Begin

Treatment of the oil and tar sludge in the base of the lagoons is more difficult. Derbyshire County Council appointed a company specialising in soil treatment technology, Celtic Technologies Ltd., to carry out treatment trials on these sludges.

Between September and November, Celtic have been working in the laboratory, carrying out initial trials. These trials have shown that these sludges can be treated by 'bioremed-iation' which breaks down oils and tars in the lagoon sludges to harmless carbon dioxide and water.

Trials on the site are now beginning. These will involve digging up 200 tonnes of the sludge and mixing it with an 'artificial soil' of colliery spoil, wood chips, fertiliser, and compost material on a treatment bed within the fenced lagoons area. The treatment bed will be turned over every so often to aerate it, which increases biological activity.

Throughout the trials the air and groundwater in the surrounding area will be monitored to make sure that their quality is not reduced by the processing. All workers on the site carry personal dose meters for volatile organic compounds, and a portable meter is used to check air quality in the country park

What is Bioremediation?

Bioremediation is a process that uses naturally occurring micro-organisms or 'bugs' (yeast, fungi, soil bacteria) to break down hydrocarbon contaminants into carbon dioxide and water. The process is similar to the garden compost heap, but much more closely controlled.

Some Questions Answered

Are the 'bugs' harmful?
Naturally occcurring indigenous soil micro-organisms are used. They are present in the soil all around us, and are not hazardous. The only difference is in the control of the process. No 'superbugs' are used!

Can bioremediation treat any contaminants?
No - only hydrocarbons (oils and tars). Asbestos cannot be broken down. Nor can toxic metals or cyanides - and these contaminants can stop bioremediation working on hydrocarbons too if the bugs find them too toxic.

Why do trials - why not go straight to a full cleanup?
Bioremediation is a new technology and needs to be checked to make sure that it will work at this site. The body paying for the reclamation, English Partnerships, wants to be sure of this before spending a lot of public money. Also by doing a small scale but closely monitored trial we can make doubly sure that there will be no adverse effects on the area before starting the full scale job.

Where do I find out more?
Not much is published in Britain, but if you're online, the United States Environmental Protection Agency publishes 'A Citizen's Guide to Bioremediation' on the Internet via their homepage on **www.epa.gov**

Access to Site

Our contractor needs to get his plant and vehicles on to the site. Country park users who park at Corbriggs are requested to keep the entrance clear by parking on the north side or in the new parking area.

Derbyshire County Council

01629 580000 ext 7173

Project Staff:
Project Director   Matt Taylor
Project Manager   Peter Storey
Project Engineer
                      Frank Westcott

A6 Summary of results of sampling the bench trials

Table A6.1 *Summary of chemical testing of selected organic contaminants in lagoon sludge and colliery spoil*

Determinant	Lagoon A1 sludge	Lagoon A sludge (mean of 2 samples)	Colliery spoil
Selected PAHs	mg/kg		
Naphthalene	77 000	27 400	1
Acenaphthylene	5400	1170	3
Acenaphthlene	156 800	24 300	5
Fluorene	81 700	9800	4
Phenanthrene	7860	4290	6
Anthracene	2240	1240	1
Fluoranthene	8250	1480	4
Pyrene	5400	1080	4
Benzo(a)anthracene	1500	200	1
Benzo(k)fluoranthene	690	110	1
Benzo(a)pyrene	1460	240	4
Indeno (1,2,3-cd)pyrene	580	94	1
Di-benzo(a,h)anthracene	140	36	< 1
Benzo(g,h,i)perylene	620	100	1
Total PAH	353 000	72 000	40
Total diesel range hydrocarbons	598 500	81 400–149 800	800
Normal alkane fraction	500 800	75 700–136 900	800
Branched/cyclic alkanes (% of DRH)	16.4	7.0–8.7	37.5
DCM extractable matter (g/kg)	1130	81.2–181.4	Not tested
Monohydric phenols	2675	198–265	Not tested

Note: Results expressed on a dry weight basis

Table A6.2 *Summary of chemical testing of selected volatile organic contaminants in lagoon sludge*

Determinant	Lagoon A1 sludge	Lagoon A sludge (mean of 2)
Benzene	3700	700
Toluene	550	250
Ethylbenzene	9.2	8.0
m/p Xylenes	95	80
o Xylene	30	26
Styrene	21	20

Note: Results expressed on a dry weight basis

Table A6.3 *Summary of chemical testing of inorganic contaminants in lagoon sludge and colliery spoil*

Determinant	Lagoon A1 sludge (mg/kg)	Lagoon A sludge (mean of 2 samples)	Colliery spoil
Total sulphate (%SO4)	0.59	1.7	1.6
Water sol. sulphate (g/l SO4)	1.5	1.6	Not tested
Sulphide	2117	211	1.4
Elemental sulphur	1490	77	500
Free cyanide	15	3	3
Complex cyanide	2325	186	340
Thiocyanate	662	501	Not tested
Available nitrogen	392	133	160
Available phosphorous	<2	<3	<1
Arsenic	55	106	131
Boron, water sol.	2.1	4.8	2.2
Cadmium	2.3	<1.5	<0.5
Chromium	39.2	14	15
Copper	98	27	75
Iron	10 980	6600	4400
Lead	137	39	65
Mercury	36	16	6
Nickel	19.6	13	30
Selenium	<1	<3	<1
Zinc	823	215	75

Note: Results expressed on a dry weight basis

Table A6.4 *Concentrations of selected contaminants – Treatment T1 (in mg/kg)*

Determinand	Time = 0 days	Time = 42 days
Naphthalene *	10 150 (5717)	33 (18)
Acenaphthlene *	12 290 (4668)	4430 (4144)
Pyrene*	443 (59)	407 (93)
Benzo(a)pyrene*	113 (11)	115 (25)
Total PAH*	33 100 (4300)	12 500 (6412)
Total diesel range hydrocarbons*	67 000 (13 190)	40 530 (9700)
Monohydric phenols	52 (4)	26 (9)
Benzene	57 (14)	26 (9)
Toluene	32 (7)	<0.01
Ethylbenzene	2.2 (0.4)	0.2 (0.3)
o Xylene	5.8 (0.7)	0.14 (0.2)
m/p Xylenes	17 (3)	0.27 (0.4)

Notes: * Expressed on a dry weight basis
All results are mean of three samples with 95 per cent confidence limits in parentheses
No chemical analyses undertaken at day 29 in Treatment T1

Table A6.5 *Concentrations of selected contaminants – Treatment T3 (in mg/kg)*

Determinand	Time = 0 days	Time = 29 days	Time = 42 days
Naphthalene *	10 710 (566)	6370 (3200)	416 (523)
Acenaphthlene *	11 773 (1820)	9800 (1640)	4960 (1680)
Pyrene*	423 (91)	266 (11)	293 (62)
Benzo(a)pyrene*	100 (10)	95 (3.5)	89 (17)
Total PAH*	32 750 (2970)	23 633 (2690)	11 610 (2970)
Total diesel range hydrocarbons*	61 570 (7710)	38 630 (2460)	36 930 (7840)
Monohydric phenols	34 (5)	21 (5)	20 (3)
Benzene	97 (37)	Not tested	<0.01
Toluene	53 (12)	Not tested	<0.01
Ethylbenzene	2.9 (0.3)	Not tested	<0.01
o Xylene	8.1 (4)	Not tested	0.05 (0.07)
m/p Xylenes	34 (10)	Not tested	0.05 (0.06)

Note * Expressed on a dry weight basis
All results are mean of three samples with 95 per cent confidence limits in parentheses

Table A6.6 *Concentrations of selected contaminants – Treatment T6 (in mg/kg)*

Determinand	Time = 0 days	Time = 29 days	Time = 42 days
Naphthalene *	10 710 (566)	4430 (972)	106 (116)
Acenaphthlene *	11 773 (1820)	5840 (700)	1710 (850)
Pyrene*	423 (91)	190 (14)	167 (40)
Benzo(a)pyrene*	100 (10)	74 (6)	64 (9)
Total PAH*	32 750 (2970)	15 293 (1,900)	5600 (1450)
Total diesel range hydrocarbons*	61 570 (7710)	29 530 (1,190)	25 000 (1690)
Monohydric Phenols	34 (5)	15 (2)	22 (10)
Benzene	97 (37)	Not tested	<0.01
Toluene	53 (12)	Not tested	<0.01
Ethylbenzene	2.9 (0.3)	Not tested	<0.01
o Xylene	8.1 (4)	Not tested	<0.01
m/p Xylenes	34 (10)	Not tested	<0.01

Note * Expressed on a dry weight basis
All results are mean of three samples with 95 per cent confidence limits in parentheses
Results from T3 have been used for T = 0

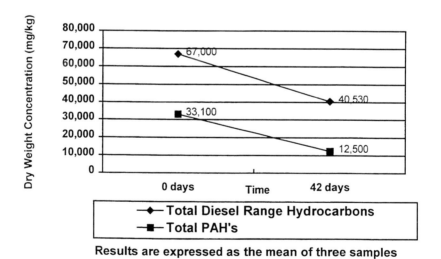

Figure A6.1 *Concentrations of selected contaminants in Treatment T1*

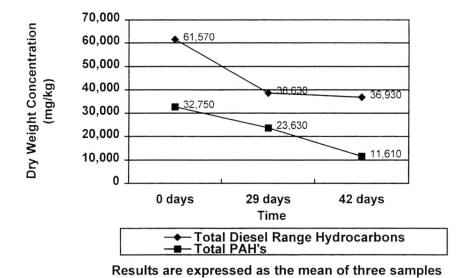

Figure A6.2 *Concentrations of selected contaminants in Treatment T3*

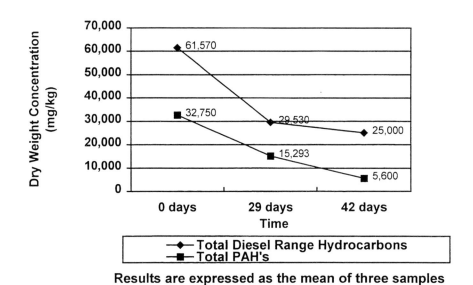

Figure A6.3 *Concentrations of selected contaminants in Treatment T6*

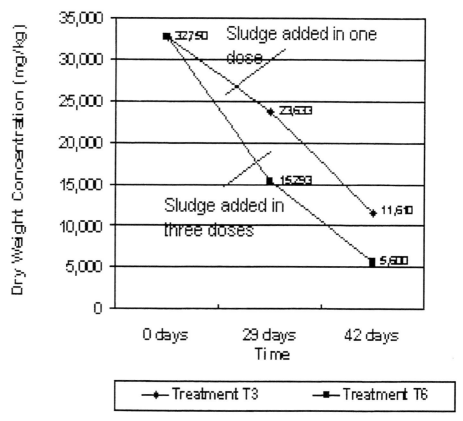

Figure A6.4 Concentrations of total PAHs in Treatment T3 and Treatment T6

Table A6.7 Summary of microcosm treatment parameters

Treatment	Parameter	Week number					
		1	2	3	4	5	6
1	VOCs	70.4	2.5	7.1	26.6	4.9	2.1
	Temperature	NT	16.8	16.9	17.1	14.2	14.3
	pH	7.4	6.5	7.8	7.3	7.3	7.3
	Moisture %	36.9	22.3	23	22	20.8	24.9
2	VOCs	65.3	3.2	6.4	30.1	5.3	3.2
	Temperature	NT	16.8	16.4	16.4	12.9	13.2
	pH	7.6	6.9	7.9	7.8	7.8	7.7
	Moisture %	35.4	29	28.3	25	24.1	23.4
3	VOCs	70.1	4.0	5.9	24.6	45	2.4
	Temperature	NT	17.1	16.5	16.6	14.3	14.3
	pH	7.4	7.3	7.6	7.7	7.4	7.3
	Moisture %	35.5	23.1	22	21	23.4	24.4
4	VOCs	26.4	3.5	6.5	20.6	4.5	1.5
	Temperature	NT	16.6	16.2	16.3	13.4	14.2
	pH	7.3	7.0	7.4	7.2	7.4	7.4
	Moisture %	23	21.3	20.1	20.5	24.2	23.6
5	VOCs	–	4.2	7.5	21.8	3.1	2.1
	Temperature	–	16.6	16.3	16.4	13.4	13.9
	pH	–	7.5	7.8	7.7	7.6	7.5
	Moisture %	–	22.1	21.7	22	20.4	23.4
6	VOCs	–	–	16.5	30	3.9	1.4
	Temperature	–	–	NT	16.2	12.9	13.5
	pH	–	–	7.8	7.5	7.4	7.4
	Moisture %	–	–	23	22.4	20.5	21

Note: NT indicates not tested – indicates not established

A7 Summary of results of sampling treatment bed (independent analysis)

Table A7.1   *Chemical contaminant testing summary – treatment beds and constituents before sludge addition (25.11.97)*

Sample number	Description	Mineral oils (mg/kg)	Monohyd. phenols (mg/kg)	Naptha-lene (mg/kg)	Acenap-thene (mg/kg)	Pyrene (mg/kg)	Benzo-a-pyrene (mg/kg)	EPA 16 PAH (mg/kg)	Mercury (mg/kg)	Lead (mg/kg)	Arsenic (mg/kg)
GML/02/01	Treatment bed A1	507	<1.5	1.5	0.74	0.66	<0.5	7.2	0.44	69.5	62.7
GML/02/02	Treatment bed B2	394	<1.5	2.1	1.6	0.82	<0.5	9.9	0.96	83.6	95.7
GML/02/03	Treatment bed A3	355	<1.5	2.8	<0.5	1.4	0.65	14	0.75	127	137
GML/02/04	Treatment bed B4	445	<1.5	3.8	2.2	1.7	0.79	19	0.88	72.1	102
GML/02/05*	Peat stockpile	707	<1.62	<0.5	<0.5	<0.5	<0.5	<5	0.12	<50	<5
GML/02/06*	Colliery spoil stockpile	142	<1.5	23	35	2.1	<0.5	657	1.38	149	93.8
	No of results	4						4	4	4	4
	Mean	425.3						12.525	0.7575	88.05	99.35
	Sample s.d.	65.8						5.142875	0.228674	26.67964	30.44717
	't'	3.182						3.182	3.182	3.182	3.182
	95% UCL	530						20.6	1.1	130.6	147.8
	95% LCL	320.6						4.4	0.5	45.6	51

* not included in means/CL's

Table A7.2 *Chemical contaminant testing summary – original treatment bed: after first sludge addition (25.11.97)*

Sample number	Description	Mineral oils (mg/kg)	Monohyd. phenols (mg/kg)	Naptha-lene (mg/kg)	Acenap-thene (mg/kg)	Pyrene (mg/kg)	Benzo-a-pyrene (mg/kg)	EPA 16 PAH (mg/kg)	Mercury (mg/kg)	Lead (mg/kg)	Arsenic (mg/kg)
GML/03/04	Treatment bed A1	10000	27.4	7900	6000	220	77	18900			
GML/03/05	Treatment bed A2	9790	8.85	4500	1600	140	23	9300			
GML/03/06	Treatment bed B2	4180	5.42	1400	750	72	25	3320			
GML/03/07	Treatment bed A3	3030	3.04	790	1300	52	15	3200			
GML/03/08	Treatment bed B3	3780	1.8	830	1600	88	17	3910			
GML/03/09	Treatment bed B4	4820	2.33	4600	4100	140	36	11600			
	No of results	6	6	6	6	6	6	6			
	Mean	5933.333	8.14	3336.667	2558.333	118.6667	32.16667	8371.667			
	Sample s.d.	3123.829	9.785824	2838.765	2045.096	61.22309	23.17254	6233.806			
	't'	2.571	2.571	2.571	2.571	2.571	2.571	2.571			
	95% UCL	9146	16.4	6256	4661	181.6	56.1	14783			
	95% LCL	2720	-0.23	417	455	55.8	8.3	1961			

Table A7.3 *Chemical contaminant testing summary – treatment bed/biopile: before second sludge addition (18.2.98)*

Sample number	Description	Mineral oils (mg/kg)	Monohyd. penols (mg/kg)	Naptha-lene (mg/kg)	Acenap-thene (mg/kg)	Pyrene (mg/kg)	Benzo-a-pyrene (mg/kg)	EPA 16 PAH (mg/kg)	Mercury (mg/kg)	Lead (mg/kg)	Arsenic (mg/kg)
GML/4/01	Treatment bed B1	4160	<1.5	440	1700	71	27	3660			
GML/4/02	Treatment bed A2	4860	<1.5	430	1800	78	26	3630			
GML/4/03	Treatment bed B3	3840	<1.5	370	2000	87	30	4090			
GML/4/04	Treatment bed B4	3860	<1.5	550	1700	77	32	3880			
GML/4/05	Mixing bed Q1	4380	<1.5	25	1800	81	28	3540			
GML/4/06	Mixing bed P2	4310	<1.5	380	1900	73	28	3880			
GML/4/07	Mixing bed Q3	4740	<1.5	300	1700	68	27	3430			
GML/4/08	Mixing bed P4	4690	<1.5	350	1900	75	28	3890			
	No of results	8		8	8	8	8	8			
	Mean	4355		355.6	1812.5	76.3	28.3	3750.0			
	Sample s.d.	390.3		152.9	112.6	6.0	1.9	219.7			
	't'	2.365		2.365	2.365	2.365	2.365	2.365			
	95% UCL	4681		483.4	1906.7	81.3	29.9	3933			
	95% LCL	4029		227.8	1718.3	71.3	26.7	3566			

Table A7.4 *Chemical contaminant testing summary – treatment bed: after second sludge addition (20.2.98)*

Sample number	Description	Mineral oils (mg/kg)	Monohyd. phenols (mg/kg)	Naptha-lene (mg/kg)	Acenap-thene (mg/kg)	Pyrene (mg/kg)	Benzo-a-pyrene (mg/kg)	EPA 16 PAH (mg/kg)	Mercury (mg/kg)	Lead (mg/kg)	Arsenic (mg/kg)
GML/5/01	Treatment bed A1	7120	2.45	2700	3100	130	42	8770			
GML/5/02	Treatment bed A1	9340	2.78	2500	3700	150	48	9320	1.39	<50	69.4
GML/5/03	Treatment bed B1	6840	2.47	4300	4900	200	66	13400			
GML/5/04	Treatment bed B1	6770	2.25	2900	3600	170	49	9810			
GML/5/05	Treatment bed A2	7520	1.8	3600	4100	160	64	11400			
GML/5/06	Treatment bed A2	8660	1.76	3900	4500	160	61	12300			
GML/5/07	Treatment bed B2	9990	2.1	4100	4600	170	65	12800			
GML/5/08	Treatment bed B2	10700	<1.5	4300	4700	180	68	13300	2.06	57.4	87.4
GML/5/09	Treatment bed A3	9300	2.77	3500	4900	210	68	12700	1.99	60.5	93.2
GML/5/10	Treatment bed A3	9790	6.4	3700	4300	170	61	11900			
GML/5/11	Treatment bed B3	8460	4.54	2900	3400	120	43	9210			
GML/5/12	Treatment bed B3	8280	4.43	3400	2600	130	64	9570			
GML/5/13	Treatment bed A4	8080	4.06	3700	2500	170	73	10000			
GML/5/14	Treatment bed A4	9480	5.42	3800	4400	170	69	12200			
GML/5/15	Treatment bed B4	10900	<1.5	3700	4100	140	49	11400	1.99	55.4	89.1
GML/5/16	Treatment bed B4	9960	5.08	4300	4600	180	71	13100			
	No of results	16		16	16	16	16	16	4	3	4
	Mean	8824.4		3581.3	4000	163.1	60.1	11323.8	1.9	57.8	84.8
	Sample s.d.	1319.2		573.0	773.7	24.7	10.3	1625.5	0.3	2.6	10.5
	't'	2.131		2.131	2.131	2.131	2.131	2.131			
	95% UCL	9527.2		3886.6	4412.2	176.3	65.6	12189.8			
	95% LCL	8121.6		3276	3587.8	149.9	54.6	10457.8			

Table A7.5 *Chemical contaminant testing summary – biopile: after second sludge addition (20.2.98)*

Sample number	Description	Mineral oils (mg/kg)	Monohyd. phenols (mg/kg)	Naptha-lene (mg/kg)	Acenap-thene (mg/kg)	Pyrene (mg/kg)	Benzo-a-pyrene (mg/kg)	EPA 16 PAH (mg/kg)	Mercury (mg/kg)	Lead (mg/kg)	Arsenic (mg/kg)
GML/05/17	Mixing bed P1	10700	<1.57	3400	3800	180	68	10800			
GML/05/18	Mixing bed P1	11300	1.88	3100	3500	170	49	10100	2.05	66.4	91.2
GML/05/19	Mixing bed Q1	9140	3.38	3400	3800	160	61	10700			
GML/05/20	Mixing bed Q1	9730	4.21	3400	4500	160	62	11700			
GML/05/21	Mixing bed P2	10500	6.9	3300	3000	150	55	9380			
GML/05/22	Mixing bed P2	9620	4	3400	3100	160	60	9800			
GML/05/23	Mixing bed Q2	11000	2.1	4200	4000	170	64	12000			
GML/05/24	Mixing bed Q2	10800	3.22	3600	3300	160	48	10200	2.57	60.4	90.4
GML/05/25	Mixing bed P3	11200	6.2	3500	3900	190	54	11200	2.27	66.1	98.7
GML/05/26	Mixing bed P3	8880	5.77	4000	4100	190	68	12100			
GML/05/27	Mixing bed Q3	11200	4.13	5100	4900	190	67	14300			
GML/05/28	Mixing bed Q3	9720	4.03	5100	5200	200	74	15000			
GML/05/29	Mixing bed P4	8440	3.51	3300	4000	190	67	11200			
GML/05/30	Mixing bed P4	9830	3.51	3300	3400	37	54	10000			
GML/05/31	Mixing bed Q4	10100	3.37	3600	4200	160	60	11500	2.05	65.7	117
GML/05/32	Mixing bed Q4	8330	2.5	2800	3400	160	45	9380			
	No of results	16		16	16	16	16	16	4	4	4
	Mean	10030.63		3656.25	3881.25	164.1875	59.75	11210	2.235	64.65	99.325
	Sample s.d.	982.7205		648.0419	615.5959	37.19179	8.282512	1601.732	0.246238	2.847806	12.36214
	't'	2.131		2.131	2.131	2.131	2.131	2.131			
	95% UCL	10554.1		4001.5	4209.3	184	64.2	12063.3			
	95% LCL	9507.1		3311.1	3553.3	144.4	55.4	9976.7			

Table A7.6 *Chemical contaminant testing summary (leach test samples) – treatment bed/biopile: after second sludge addition (20.2.98)*

Sample number	Description	EPA 16 PAH (mg/l)	Monohyd. phenols (mg/l)	Ammoniac. nitrogen (mg/l)	Cyanide total (mg/l)	COD (settled) (mg/l)	Sulphate (mg/l)	Lead (mg/l)	Arsenic (ug/l)
GML/05/02	Treatment bed A1	6.36	0.17	398	1.11	903	602	< 0.01	240
GML/05/08	Treatment bed B2	4.59	0.19		1.19		514	< 0.01	1030
GML/05/09	Treatment bed A3	5.58	0.19	312	1.16	679	629	< 0.01	250
GML/05/15	Treatment bed B4	5.51	0.22		1.23		527	< 0.01	1040
GML/05/18	Mixing bed P1	3.62	0.39		1.38		< 1.00	< 0.01	220
GML/05/24	Mixing bed Q2	3.67	0.33	253	1.69	704	725	< 0.01	29
GML/05/25	Mixing bed P3	7.58	0.41		1.25		< 1.00	< 0.01	140
GML/05/31	Mixing bed Q4	5.05	0.28	391	1.6	769	646	< 0.01	260
	No of results	8		4	8	4			8
	Mean	5.2		338.5	1.3	763.8			401.1
	Sample s.d.	1.3		69.1	0.2	100.3			398.5
	't'	2.365		3.182	2.365	3.182			2.365
	95% UCL	6.3		448.4	1.5	923.4			734.3
	95% LCL	4.1		237.4	1.1	604.2			67.9

Table A7.7 *Chemical contaminant testing summary – treatment bed: four weeks after second sludge addition (27.3.98)*

Sample number	Description	Mineral oils (mg/kg)	Monohyd. phenols (mg/kg)	Naptha-lene (mg/kg)	Acenap-thene (mg/kg)	Pyrene (mg/kg)	Benzo-a-pyrene (mg/kg)	EPA 16 PAH (mg/kg)	Mercury (mg/kg)	Lead (mg/kg)	Arsenic (mg/kg)
GML/07/01	Treatment bed A1	6130	1.87	31	2600	150	49	5100			
GML/07/02	Treatment bed B1	5200	1.86	62	2900	160	44	5640			
GML/07/03	Treatment bed A2	5750	2.39	540	3500	170	61	7430			
GML/07/04	Treatment bed B2	5800	2.09	44	2500	130	46	5000			
GML/07/05	Treatment bed A3	6020	2.19	300	3300	160	60	6750			
GML/07/06	Treatment bed B3	6010	2.41	520	3500	160	60	7250			
GML/07/07	Treatment bed A4	6990	2.4	380	2900	140	51	6120			
GML/07/08	Treatment bed B4	6120	2.46	350	3400	170	64	7110			
	No of results	8		8	8	8	8	8			
	Mean	6002.5		278.4	3075.0	155.0	54.4	6300.0			
	Sample s.d.	500.6		209.0	402.7	14.1	7.7	973.7			
	't'	2.365		2.365	2.365	2.365	2.365	2.365			
	95% UCL	6421		453.2	3412	166.8	55.2	7114			
	95% LCL	5584		103.64	2738	143.2	53.6	5486			

Table A7.8 *Chemical contaminant testing summary (leach test samples) – treatment bed: four weeks after second sludge addition (27.3.98)*

Sample number	Description	EPA 16 PAH (mg/l)	Monohyd. phenols (mg/l)	Ammoniac. nitrogen (mg/l)	Cyanide total (mg/l)	COD (settled) (mg/l)	Sulphate (mg/l)	Lead (mg/l)	Arsenic (ug/l)
GML/07/03	Treatment bed A2	1.57	< 0.15		0.85	850	654	0.01	0.19
GML/07/06	Treatment bed B3	2.08	< 0.15		0.77	840	918	< 0.01	0.27
	No of results	2			2	2	2		2
	Mean	1.8			0.8	845.0	786.0		0.2
	Sample s.d.	0.4			0.1	7.1	186.7		0.1
	't'								
	95% UCL								
	95% LCL								

Table A7.9 *Chemical contaminant testing summary – treatment bed: seven weeks after second sludge addition (15.4.98)*

Sample number	Description	Mineral oils (mg/kg)	Monohyd. phenols (mg/kg)	Naptha-lene (mg/kg)	Acenap-thene (mg/kg)	Pyrene (mg/kg)	Benzo-a-pyrene (mg/kg)	EPA 16 PAH (mg/kg)	Mercury (mg/kg)	Lead (mg/kg)	Arsenic (mg/kg)
GML/08/01	Treatment bed A1	7500	< 1.50	36	< 3000	150	37	1750			
GML/08/02	Treatment bed A1	7540	< 1.50	130	2900	130	53	5520	2.37	73.4	100
GML/08/03	Treatment bed B1	7840	< 1.50	41	< 2300	150	47	1950			
GML/08/04	Treatment bed B1	7830	< 1.50	170	< 3000	160	50	2290			
GML/08/05	Treatment bed A2	8030	< 1.50	75	< 2900	150	49	2010			
GML/08/06	Treatment bed A2	6860	< 1.50	74	3000	150	45	5140			
GML/08/07	Treatment bed B2	6670	< 1.50	100	< 1900	130	42	1870			
GML/08/08	Treatment bed B2	6720	< 1.50	110	2400	110	42	4420	1.42	51.3	68
GML/08/09	Treatment bed A3	8240	< 1.50	140	< 2700	130	44	1920	2.01	< 50.0	< 5.0
GML/08/10	Treatment bed A3	6410	< 1.50	150	< 1330	66	26	1230			
GML/08/11	Treatment bed B3	6830	< 1.50	95	2600	110	44	4890			
GML/08/12	Treatment bed B3	6380	< 1.50	50	3000	150	56	5410			
GML/08/13	Treatment bed A4	6480	< 1.50	67	2900	150	49	5440			
GML/08/14	Treatment bed A4	7740	1.51	180	2900	130	43	5640			
GML/08/15	Treatment bed B4	7190	< 1.50	110	< 2600	110	42	1970			
GML/08/16	Treatment bed B4	7150	< 1.50	55	< 2300	190	50	2230			
	No of results	16		16		16	16	16			
	Mean	7213.1		98.9		135.4	44.9	3355.0			
	Sample s.d.	613.9		45.5		28.0	7.0	1724.6			
	't'	2.131		2.131		2.131	2.131	2.131			
	95% UCL	7700		123.1		150.3	48.6	4273.8			
	95% LCL	6726.2		74.7		120.5	41.2	2436.2			

Table A7.10 *Chemical contaminant testing summary (leach test results) – treatment bed: leach test results (15.4.98)*

Sample number	Description	EPA 16 PAH* (mg/l)	Monohyd. phenols (mg/l)	Ammoniac. nitrogen (mg/l)	Cyanide total (mg/l)	COD (settled) (mg/l)	Sulphate (mg/l)	Lead (mg/l)	Arsenic* (ug/l)
GML/08/02	Treatment bed	3.32	< 0.15	192	1.01	1420	500	< 0.05	195
GML/08/08	Treatment bed	2.85	< 0.15		0.93		3880	< 0.05	1120
GML/08/09	Treatment bed	2.23	< 0.15	208	0.76	1210	645	< 0.05	192
GML/08/15	Treatment bed	2.79	< 0.15		1.04		7710	0.36	3200
	No of results	4		2	4	2	4		
	Mean	2.8		200.0	0.9	1315.0	3183.8		
	Sample s.d.	0.4		11.3	0.1	148.5	3397.0		
	't'								
	95% UCL								
	95% LCL								

* (EPA 16 PAH and Arsenic converted to mg and ug respectively)

Table A7.11 *Chemical contaminant testing summary – biopile: seven weeks after second sludge addition (15.4.98)*

Sample number	Description	Mineral oils (mg/kg)	Monohyd. phenols (mg/kg)	Naptha-lene (mg/kg)	Acenap-thene (mg/kg)	Pyrene (mg/kg)	Benzo-a-pyrene (mg/kg)	EPA 16 PAH (mg/kg)	Mercury (mg/kg)	Lead (mg/kg)	Arsenic (mg/kg)
GML/08/17	Biopile P1	9770	1.86	1400	< 4100	200	66	4830			
GML/08/18	Biopile P1	4430	< 1.50	1500	2300	99	30	5860	1.41	56.4	85.1
GML/08/19	Biopile Q1	12 400	1.82	3800	< 3200	320	69	7740			
GML/08/20	Biopile Q1	11 100	2.66	4300	4500	200	69	12 900			
GML/08/21	Biopile P2	11 300	< 1.50	1100	2700	130	47	6370			
GML/08/22	Biopile P2	9120	1.91	2000	< 2700	200	54	5100			
GML/08/23	Biopile Q2	13 800	2.99	3400	2400	180	60	6370			
GML/08/24	Biopile Q2	13 600	< 1.50	3100	< 4500	210	67	6710	2.6	67	87.9
GML/08/25	Biopile P3	7500	< 1.50	1100	3100	170	62	7600	1.65	51.1	72
GML/08/26	Biopile P3	12 800	3.48	2900	2900	130	50	8830			
GML/08/27	Biopile Q3	11 900	1.86	1900	2800	160	59	7930			
GML/08/28	Biopile Q3	13 400	3.05	2300	3900	200	67	10 000			
GML/08/29	Biopile P4	12 000	< 1.50	640	< 3000	110	33	3160			
GML/08/30	Biopile P4	12 000	3.53	3200	3200	170	63	9850			
GML/08/31	Biopile Q4	12 000	< 1.50	830	2000	91	36	4840	2.18	67.2	84.6
GML/08/32	Biopile Q4	11 800	1.76	2600	< 3100	120	36	5220			
	No of results	16		16		16	16	16			
	Mean	11182.5		2254.4		168.1	54.3	7081.9			
	Sample s.d.	2450.0		1120.2		56.8	13.8	2449.8			
	't'	2.131		2.131		2.131	2.131	2.131			
	95% UCL	12487.7		2851.2		198.4	61.7	8387			
	95% LCL	9877.3		1657.6		137.8	46.9	5776.8			

Table A7.12 *Chemical contaminant testing summary (leach test sampling) – biopile: leachate test results (15.4.98)*

Sample number	Description	EPA 16 PAH* (mg/l)	Monohyd. phenols (mg/l)	Ammoniac. nitrogen (mg/l)	Cyanide total (mg/l)	COD (settled) (mg/l)	Sulphate (mg/l)	Lead (mg/l)	Arsenic* (ug/l)
GML/08/18	Biopile	3.63	< 0.15		0.9		735	< 0.05	400
GML/08/24	Biopile	6.01	< 0.15	260	1.13	1400	630	< 0.05	313
GML/08/25	Biopile	3.53	< 0.15		0.86		435	< 0.05	360
GML/08/31	Biopile	6.18	< 0.15	242	0.91	600	765	< 0.05	301
	No of results	4		2	4	2	4		
	Mean	4.8		251.0	1.0	1000.0	641.3		
	Sample s.d.	1.5		12.7	0.1	565.7	149.2		
	't'								
	95% UCL								
	95% LCL								

* (EPA 16 PAH and Arsenic converted to mg and ug respectively)

Table A7.13 *Chemical contaminant testing summary – original treatment bed: at end of trials (15.4.98)*

Sample number	Description	Mineral oils (mg/kg)	Monohyd. phenols (mg/kg)	Naptha-lene (mg/kg)	Acenap-thene (mg/kg)	Pyrene (mg/kg)	Benzo-a-pyrene (mg/kg)	EPA 16 PAH (mg/kg)	Mercury (mg/kg)	Lead (mg/kg)	Arsenic (mg/kg)
GML/08/33	Original treatment bed A1	7100	1.82	390	2600	100	41	4510			
GML/08/34	Original treatment bed A1	6670	< 1.67	690	2500	90	36	4790	1.59	69.8	113
GML/08/35	Original treatment bed B1	5800	< 1.50	140	< 2400	100	34	1270			
GML/08/36	Original treatment bed B1	1570	< 1.50	31	400	40	13	762			
	No of results	4		4		4	4	4			
	Mean	5285.0		312.8		82.5	31.0	2833.0			
	Sample s.d.	2535.0		293.0		28.7	12.4	2111.4			
	't'	3.182		3.182		3.182	3.182	3.182			
	95% UCL	9318		779		128.2	50.7	6192.2			
	95% LCL	1252		<0		36.8	11.3	< 0			

Table A7.14 *Chemical contaminant testing summary – temporary stockpile remainder: at end of trials (15.4.98)*

Sample number	Description	Mineral oils (mg/kg)	Monohyd. phenols (mg/kg)	Napthalene (mg/kg)	Acenapthene (mg/kg)	Pyrene (mg/kg)	Benzo-a-pyrene (mg/kg)	EPA 16 PAH (mg/kg)	Mercury (mg/kg)	Lead (mg/kg)	Arsenic (mg/kg)
GML/08/37	Temp. stockpile remainder	4370	< 1.50	150	< 2200	59	19	1250	1.57	65	92.2
GML/08/38	Temp. stockpile remainder	3070	< 1.50	13	770	56	23	1150			
GML/08/39	Temp. stockpile remainder	3370	< 1.50	8.5	< 215	94	40	454			
GML/08/40	Temp. stockpile remainder	3030	< 1.50	14	< 550	44	18	334			
	No of results	4		4		4	4	4			
	Mean	3460.0		46.4		63.3	25.0	797.0			
	Sample s.d.	625.3		69.1		21.5	10.2	469.7			
	't'	3.182		3.182		3.182	3.182	3.182			
	95% UCL	4455		156.3		97.5	41.2	1544.3			
	95% LCL	2465		<0		29.1	8.8	49.7			

Table A7.15 *Chemical contaminant testing summary (leachate samples): leachate test results – old treatment bed and stockpile remainder (15.4.98)*

Sample number	Description	EPA 16 PAH* (mg/l)	Monohyd. phenols (mg/l)	Ammoniac. nitrogen (mg/l)	Cyanide total (mg/l)	COD (settled) (mg/l)	Sulphate (mg/l)	Lead (mg/l)	Arsenic* (ug/l)
GML/08/34	Old treatment bed	1.21	< 0.15	52.8	0.29	185	640	< 0.05	75
GML/08/37	Stockpile Remainder	0.28	< 0.15	122	< 0.20	116	1140	< 0.05	75
	No of results								
	Mean								
	Sample s.d.								
	't'								
	95% UCL								
	95% LCL								

* (EPA 16 PAH and Arsenic converted to mg and ug respectively)

Table A7.16 *Chemical contaminant testing summary – water sample test results (16.9.97)*

Sample number	Description	EPA 16 PAH (mg/l)	Monohyd. phenols (mg/l)	Ammoniac nitrogen (mg/l)	Cyanide total (mg/l)	COD (settled) (mg/l)	Sulphate (mg/l)	Lead (mg/l)	Arsenic (mg/kg)
GML/1/03	Effluent flowmeter	0.01	<0.15	93	7.21	370	1590	<0.10	2.74
GML/1/04	Lagoon C	0.26	1.06	103	12.3	448	2020	<0.10	5.4
GML/1/05	BH 9			145	1.69	407			
GML/1/06	BH 12			104	1.57	284			
GML/1/07	BH M3			19	<0.20	94			
GML/2/07	Lagoon C			88.9	27.1	265			
GML/2/08	Effluent flowmeter			84.2	20.6	220			
GML/2/09	BH 9			42.2	0.74	145			
GML/3/01	Lagoon A	1.55	5.39	5.25	2.45	859	378	<0.10	9.61
GML/3/02	BH 15			112	182	782			
GML/3/03	Bh 3			94.2	141	685			
GML/4/09	Return pump discharge			82.6	10.5	582			
GML/4/10	Lagoon C			43.4	12.8	497			
GML/6/01	Effluent Flowmeter	<10.0	<0.15	106	11.7	350	1750	<0.10	41.6
GML/6/02	BH 4	19.7	<0.15	5.66	<0.20	150	38.2	<0.10	32.2
GML/6/03	BH 3	1140	0.18	110	21.4	1020	943	<0.10	739
GML/6/04	BH M2	<10.0	<0.15	33.2	0.37	7660	1470	0.13	48.2
GML/6/05	BH M3	<10.0	<0.15	17.6	<0.20	1890	654	0.11	279
GML/6/06	BH M9	<0.10	<0.15	24.8	0.48	369	1650	0.19	846
GML/9/03	BH 3								
GML/9/04	BH 15								
GML/9/05	BH 4								
GML/9/06	BH 9								
GML/9/07	BH M9								
GML/9/08	BH M3								
GML/9/09	Effluent flowmeter								

A8 Summary of results of treatment bed monitoring

Table A8.1 *Bed 1 monitoring: total concentration of VOCs (ppm)*

Date	Location 1		Location 2		Location 3		Average	
	10 cm	Ambient	10 cm	Ambient	10 cm	Ambient	10 cm	Ambient
7.01.98	1.2	1	2.4	1.0	3.7	1.0	2.4	1.0
8.01.98	1.6	0.9	1.5	1.2	1.3	1.2	1.5	1.1
12.01.98	10.3	4.3	10.1	5	10	3.4	10.1	4.2
13.01.98	1.5	1.3	2.5	1.7	3.0	1.3	2.3	1.4
14.01.98	8.0	5.2	7.1	4.5	4.9	5.0	6.7	4.9
15.01.98	1.3	1.2	1.5	1.0	2.1	0.8	1.6	1.0
16.01.98	2.5	1.5	3.7	2.0	2.5	2.0	2.9	1.8
19.01.98	11	5.0	10.2	4.1	5.9	4.5	9	4.5
20.01.98	1.6	0.7	1.4	1.3	1.4	1.2	1.5	1.1
21.01.98	1.3	0.9	1.5	0.9	1.5	0.9	1.4	0.9
23.01.98	1.5	1.0	1.5	1.0	2	1.0	1.7	1.0
26.01.98	2.3	1.1	3.9	1.2	5.9	1.1	4.0	1.1
27.01.98	3.2	1.4	5.7	1.4	4.3	0.9	4.4	1.2
28.01.98	4.5	1.3	3.1	1.1	2.9	1.5	3.5	1.3

Note: Bed 1 used to build biopile on 29 January 1998

Table A8.2 *Bed 1 monitoring: pH*

Sample	Date			
	9.01.98	16.01.98	23.01.98	30.01.98
Bed 1.location 1	8.2	8.8	8.5	8.6
Bed 1.location 2	8.2	8.7	8.6	9.1
Bed 1.location 3	8.1	8.9	8.7	8.9
Mean	8.2	8.8	8.6	8.9

Table A8.3 *Bed 1 monitoring: moisture content (percentage)*

Sample	Date			
	9.01.98	16.01.98	23.01.98	30.01.98
Bed 1.location 1	38.8	45.7	61.4	40.1
Bed 1.location 2	35.0	38.0	33.8	37.4
Bed 1.location 3	41.9	45.7	37.1	60.5
Mean	38.6	43.1	44.1	46.0

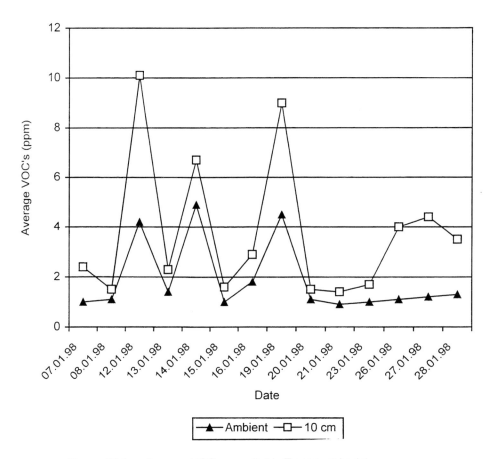

Figure A8.1   *Average VOCs recorded in Treatment bed 1*

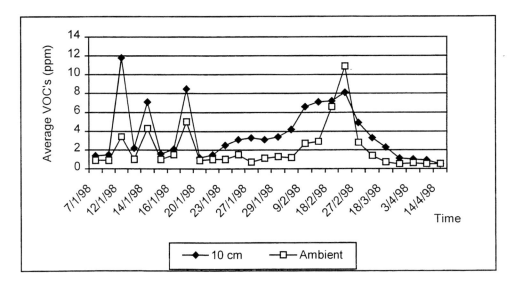

Figure A8.2   *Average VOCs recorded in Treatment bed 2*

Table A8.4 *Bed 2 monitoring: total concentration of VOCs (ppm)*

DATE	Location 1		Location 2		Location 3		Average	
	10 cm	Ambient	10 cm	Ambient	10 cm	Ambient	10 cm	Ambient
7.01.98	1.5	0.6	1.4	0.5	1.3	1.5	1.4	0.9
8.01.98	1.5	1.0	1.6	0.9	1.3	0.9	1.5	0.9
12.01.98	12.8	3.5	12.4	2.3	10.1	4.3	11.8	3.4
13.01.98	2.0	1.2	1.5	0.9	3.0	0.8	2.2	1.0
14.01.98	6.2	3.5	7.0	4.3	8.1	5.0	7.1	4.3
15.01.98	1.2	0.8	1.7	1.2	2.0	1.0	1.6	1.0
16.01.98	2.1	1.5	2.1	1.6	2.1	1.5	2.1	1.5
19.01.98	8.5	4.3	9.2	5.0	7.8	5.7	8.5	5.0
20.01.98	1.2	0.9	1.1	0.9	1.2	0.9	1.2	0.9
21.01.98	1.5	1.0	1.5	1.0	1.5	1.0	1.5	1.0
23.01.98	2.0	1.0	2.4	1.1	3.2	1.0	2.5	1.0
26.01.98	3.0	1.5	3.0	1.5	3.2	1.5	3.1	1.5
27.01.98	2.4	0.6	6.4	0.5	1.2	0.9	3.3	0.7
28.01.98	1.9	1.1	5.8	1.2	1.5	1.0	3.1	1.1
29.01.98	3.5	1.3	3.6	1.5	3.2	1.0	3.4	1.3
30.01.98	3.5	1.1	2.1	1.4	6.9	1.2	4.2	1.2
9.02.98	5.8	2.5	6.4	3.1	7.5	2.6	6.6	2.7
10.02.98	5.3	3.1	7.8	2.9	8.1	2.7	7.1	2.9
18.02.98	7.0	6.9	6.7	6.3	8.0	6.7	7.2	6.6
20.02.98	7.4	9.1	8.3	11.6	8.7	12	8.1	10.9
27.02.98	5.3	2.4	5.1	3.0	4.3	2.9	4.9	2.8
10.03.98	3.4	1.5	3.4	1.9	3.2	0.9	3.3	1.4
18.03.98	1.8	0.8	2.1	0.6	2.9	0.7	2.3	0.7
27.03.98	1.1	0.5	0.9	0.5	1.4	0.5	1.1	0.5
3.04.98	1.0	0.6	0.8	0.6	1.2	0.5	1.0	0.6
9.04.98	0.9	0.5	0.7	0.5	1.1	0.6	0.9	0.5
14.04.98	0.9	0.5	0.7	0.5	0.9	0.5	0.8	0.5

Table A8.5 *Bed 3 monitoring: total concentration of VOCs (ppm)*

DATE	Location 1		Location 2		Location 3		Average	
	10 cm	Ambient	10 cm	Ambient	10 cm	Ambient	10 cm	Ambient
7.01.98	1.3	1.2	13	1.3	3.6	1.1	6.0	1.2
8.01.98	1.2	0.5	1.3	0.6	1.2	0.9	1.2	0.7
9.01.98	–	–	–	–	–	–	–	–
12.01.98	10	3.5	7.1	5.0	8.5	5.0	8.5	4.5
13.01.98	3.0	0.9	2.6	1.0	2.0	1.2	2.5	1.0
14.01.98	8.2	5	9.3	4.8	15.1	4.9	10.9	4.9
15.01.98	1.1	0.9	1.4	0.9	2.5	1.1	1.7	1.0
16.01.98	2.6	1.3	2.4	1.6	2.7	1.4	2.6	1.4
19.01.98	7.7	5.3	8.9	4.9	10	4.5	8.9	4.9
20.01.98	1.1	0.9	1.2	1.0	1.1	1.0	1.1	1.0
21.01.98	1.3	1.1	1.2	1.0	1.2	1.0	1.2	1.0
23.01.98	2.4	1.0	1.9	1.1	1.7	1.1	2.0	1.1
26.01.98	2.5	1.1	2.0	1.3	4.5	1.3	3.0	1.2
27.01.98	4.2	2.4	2.1	1.5	3.4	1.5	3.2	1.8
28.01.98	2.6	1.3	3.2	1.3	2.9	1.3	2.9	1.3

Note: Bed 3 used to build the biopile on 29 January 1998

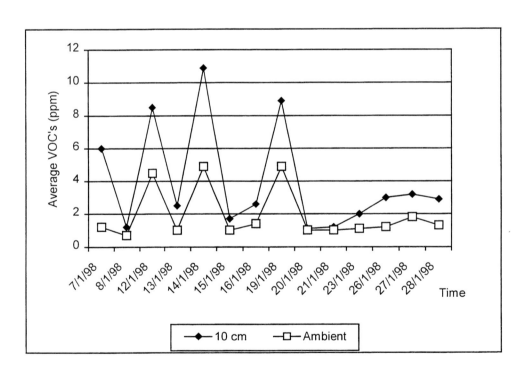

Figure A8.3 *Average VOCs recorded in Treatment Bed 3*

Table A8.6 *Bed 3 monitoring: pH*

Sample	Date			
	9.01.98	**16.01.98**	**23.01.98**	**30.01.98**
Bed 3.location 1	7.6	9.2	9.1	9.3
Bed 3.location 2	8.7	9.2	8.9	9.3
Bed 3.location 3	8.7	9.3	8.9	9.2
Mean	8.3	9.2	9.0	9.3

Table A8.7 *Bed 3 monitoring: moisture content (percentage)*

Sample	Date			
	9.01.98	**16.01.98**	**23.01.98**	**30.01.98**
Bed 3.location 1	39.7	41.2	38.2	40.3
Bed 3.location 2	29.2	41.1	37.7	38.7
Bed 3.location 3	40.0	45.0	39.3	38.2
Mean	36.3	42.4	38.4	39.1

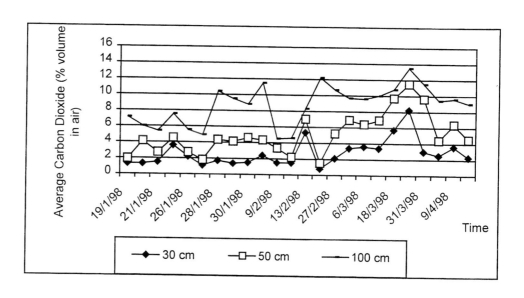

Figure A8.4 *Average carbon dioxide levels recorded in the stockpile*

Table A8.8 *Stockpile monitoring: carbon dioxide (percentage volume in air)*

Date	Location 1			Location 2			Location 3			Average		
	30 cm	50 cm	100 cm	30 cm	50 cm	100 cm	30 cm	50 cm	100 cm	30 cm	50 cm	100 cm
19.01.98	1.2	2.7	17.3	1.1	1.2	1.3	1.6	1.9	2.7	1.3	1.9	7.1
20.01.98	0.3	7.1	10.4	3.0	4.5	6.8	0.5	0.6	0.8	1.3	4.1	6.0
21.01.98	0.4	2.2	3.5	1.9	3.0	6.1	2.2	3	6.5	1.5	2.7	5.4
23.01.98	0.7	3.2	7.9	5.1	5.1	8.5	4.9	5.1	6	3.6	4.5	7.5
26.01.98	2.7	2.8	4.5	3.7	4.5	11.1	0.2	0.7	0.8	2.2	2.7	5.5
27.01.98	1.2	2.2	11.0	1.8	2.3	2.2	0.2	0.8	1.4	1.1	1.8	4.9
28.01.98	2.7	5.6	8.0	0.7	5.6	17.1	1.6	1.6	6.0	1.7	4.3	10.4
29.01.98	1.3	4.6	8.1	1.2	6.0	12.4	1.6	1.7	8.0	1.4	4.1	9.5
30.01.98	1.4	5.0	8.1	1.6	7.0	10.3	1.6	1.7	8.1	1.5	4.6	8.8
6.02.98	2.6	4.9	12	1.3	2.6	14.2	3.2	5.5	8.3	2.4	4.3	11.5
9.02.98	2.8	6.0	8.0	0.3	0.4	0.6	1.4	3.6	5.0	1.5	3.3	4.5
10.02.98	4.0	4.0	8.9	0.2	1.2	2.8	0.3	1.3	2.2	1.5	2.2	4.6
13.02.98	0.2	1.6	3.3	0.8	0.8	1.2	14.8	18.5	20.4	5.3	7.0	8.3
25.02.98	0.1	1.8	12.2	0.5	0.8	8.8	1.7	1.9	15.6	0.8	1.5	12.2
27.02.98	2.2	7.7	12.8	0.2	0.3	1.9	3.8	7.5	17.3	2.1	5.2	10.7
4.03.98	1.6	5.8	13.7	0	0	0.2	8.6	14.8	15.2	3.4	6.9	9.7
6.03.98	2.1	4.1	13.0	0	0.1	0.3	8.6	14.9	15.5	3.6	6.4	9.6
10.03.98	1.6	5.8	13.7	0.1	0.4	1.2	8.6	14.6	15.2	3.4	6.9	10.0
18.03.98	5.0	14.4	15.0	2.6	3.6	5.2	9.6	11.1	12.3	5.7	9.7	10.8
27.03.98	8.0	12.1	14.7	5.7	6.6	8.4	10.8	15.8	17.5	8.2	11.5	13.5
31.03.98	8.4	14.6	16.4	0.2	9.7	11.2	0.3	4.5	6.9	3.0	9.6	11.5
3.04.98	1.5	2.1	12.7	5.7	7.1	8.7	0.3	4.0	6.9	2.5	4.4	9.4
9.04.98	2.1	4.1	13.0	0	0.1	0.3	8.6	14.9	15.5	3.6	6.4	9.6
14.04.98	1.5	2.5	12.3	5.1	7.0	8.3	0.3	4.0	6.4	2.3	4.5	9.0

Note: All ambient levels were recorded at nil.

Table A8.9 *Stockpile monitoring: total concentration of VOCs (ppm)*

Date	Location 1		Location 2		Location 3		Average	
	10 cm depth	Ambient	10 cm depth	Ambient	10 cm depth	Ambient	10 cm depth	Ambient
19.01.98	2.0	1.9	2.3	2.0	2.1	1.9	2.1	1.9
20.01.98	2.6	1.5	1.8	2.1	2.3	1.5	1.9	1.7
23.01.98	2.1	1.6	2.4	1.5	3.0	1.3	2.5	1.5
26.01.98	3.0	1.1	3.1	3.0	2.1	2.3	2.7	2.1
27.01.98	4.3	1.0	3.9	1.2	3.2	2.1	3.8	1.4
28.01.98	2.6	1.4	2.5	1.1	2.5	3.4	2.5	2.0
29.01.98	4.6	1.2	3.1	1.1	2.6	3.6	3.4	2.0.
30.01.98	3.1	1.5	2.1	1.3	2.3	2.1	2.5	1.6.
6.02.98	12.3	3.1	13.2	2.5	12.1	3.5	12.5	3.0
9.02.98	7.5	3.2	4.9	3.1	11.0	3.2	7.8	3.2
10.02.98	6.9	2.7	7.1	2.7	5.7	2.7	6.6	2.7
25.02.98	1.9	1.2	3.1	1.2	2.9	1.2	2.6	1.2
27.02.98	1.4	0.9	2.9	0.9	3.1	0.9	2.5	0.9
10.03.98	1.5	1.1	2.6	1.1	2.8	1.1	2.3	1.1
18.03.98	1.2	1.5	1.6	1.5	2.7	1.5	1.8	1.5
27.03.98	1.3	0.9	2.5	0.9	2.8	0.9	2.2	0.9
3.04.98	1.9	1.2	2.0	1.2	1.5	1.0	1.8	1.1
9.04.98	1.4	0.9	1.9	0.9	1.1	1.0	1.5	0.9
14.04.98	1.5	1.1	0.8	1.1	1.1	1.0	1.1	1.1

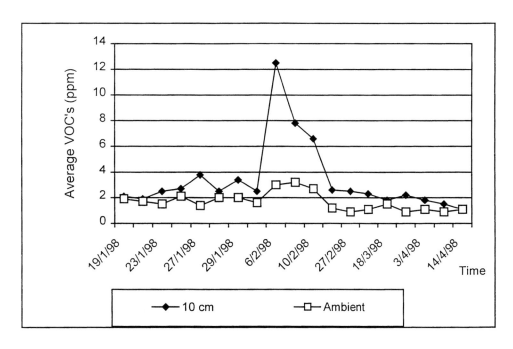

Figure A8.5 *Average VOCs measured in the stockpile*

A9 Summary of results of sampling treatment bed after completion of the trials

Table A9.1 *Chemical contaminant testing summary (leach test samples) – treatment bed: 16 weeks after second sludge addition and nine weeks after end of trials (12.6.98)*

Sample number	Description	EPA 16 PAH (mg/l)	Monohyd. phenols (mg/l)	Ammoniac. nitrogen (mg/l)	Cyanide total (mg/l)	COD (settled) (mg/l)	Sulphate (mg/l)	Lead (mg/l)	Arsenic (mg/l)
GML/09/01	Treatment bed A1	1.08		166	0.43	680	367	<0.01	0.139
GML/09/08	Treatment bed B2	0.42			0.43		257	<0.01	0.108
GML/09/11	Treatment bed B3			95.8	0.42	380	197	<0.01	0.099
GML/09/13	Treatment bed A4	1.73			0.45		220	<0.01	0.099
	No of results	3			4		4		4
	Mean	1.08			0.43		260.0		0.1
	Sample s.d.	0.66			0.01		75.3		0.02
	't'								
	95% UCL								
	95% LCL								

Table A9.2 *Chemical contaminant testing summary – treatment bed: 16 weeks after second sludge addition and nine weeks after end of trials (12.6.98)*

Sample number	Description	Mineral oils (mg/kg)	Monohyd. phenols (mg/kg)	Naptha-lene (mg/kg)	Acenap-thene (mg/kg)	Pyrene (mg/kg)	Benzo-a-pyrene (mg/kg)	EPA 16 PAH (mg/kg)	Mercury (mg/kg)	Lead (mg/kg)	Arsenic (mg/kg)
GML/09/01	Treatment bed A1	5190	<1.50	34	1300	160	65	2970	1.67	65.2	93
GML/09/02	Treatment bed A1	1080	<1.50	15	350	120	40	1050			
GML/09/03	Treatment bed B1	3310	<1.50	27	800	150	61	1910			
GML/09/04	Treatment bed B1	3020	<1.50	29	140	120	40	763			
GML/09/05	Treatment bed A2	3070	<1.50	34	820	130	40	2270			
GML/09/06	Treatment bed A2	4250	<1.50	46	990	180	51	2620			
GML/09/07	Treatment bed B2	3560	<1.50	31	660	170	57	1670			
GML/09/08	Treatment bed B2	4250	<1.50	26	730	140	54	1840	1.71	58.6	90.6
GML/09/09	Treatment bed A3	3540	<1.50	37	560	180	55	2030			
GML/09/10	Treatment bed A3	3250	<1.50	36	410	150	48	1310			
GML/09/11	Treatment bed B3	4690		28	710	120	39	1850	1.69	63.8	126
GML/09/12	Treatment bed B3	4260	<1.50	33	610	170	65	1830			
GML/09/13	Treatment bed A4	1650	<1.50	29	770	160	58	2100	1.5	64.7	94.4
GML/09/14	Treatment bed A4	3750	<1.50	30	930	130	41	2360			
GML/09/15	Treatment bed B4	3560	<1.50	42	460	170	51	1470			
GML/09/16	Treatment bed B4	3470	<1.50	30	280	160	56	1020			
	No of results	16		16	16	16	16	16	4	4	4
	Mean	3493.8		31.7	657.5	150.6	51.3	1816.4	1.64	63.1	101.0
	Sample s.d.	1027.6		7.0	292.3	21.4	9.1	596.9	0.1	3.0	16.7
	't'										
	95% UCL										
	95% LCL										

Table A9.3   *Chemical contaminant testing summary – treatment bed:*
26 weeks after second sludge addition and 19 weeks after end of trials (27.8.98)

Sample number	Description	Mineral oils (mg/kg)	Monohyd. phenols (mg/kg)	Naptha-lene (mg/kg)	Acenap-thene (mg/kg)	Pyrene (mg/kg)	Benzo-a-pyrene (mg/kg)	EPA 16 PAH (mg/kg)	Mercury (mg/kg)	Lead (mg/kg)	Arsenic (mg/kg)
GML/10/01	Treatment bed A1	2580	1.69	53	252	135	32.3	1030			
GML/10/02	Treatment bed A1	2630	4.85	56.4	271	158	43.4	1190			
GML/10/03	Treatment bed B1	3020	2.93	60.2	826	186	206	3530			
GML/10/04	Treatment bed B1	1810	2.3	45.7	123	132	36.4	791			
GML/10/05	Treatment bed A2	1890	4.52	58.5	106	141	39.9	812			
GML/10/06	Treatment bed A2	2120	1.73	57.5	171	135	37.5	902			
GML/10/07	Treatment bed B2	2260	2.03	50.2	154	138	36.2	909			
GML/10/08	Treatment bed B2	2640	3.64	52.5	277	139	39.8	1120			
GML/10/09	Treatment bed A3	2170	<1.5	49.7	156	141	36.7	899			
GML/10/10	Treatment bed A3	2700	<1.5	41.2	445	128	33.3	1380			
GML/10/11	Treatment bed B3	3480	<1.5	53.7	504	129	40.8	1420			
GML/10/12	Treatment bed B3	2400	1.97	50.3	210	148	40.8	977			
GML/10/13	Treatment bed A4	2070	3.69	41.7	262	118	33.8	986			
GML/10/14	Treatment bed A4	1920	4.43	48.9	119	121	35.1	746			
GML/10/15	Treatment bed B4	2600	1.9	59	30	39.9	35.8	530			
GML/10/16	Treatment bed B4	1770	1.7	52.1	29	8.14	32	445			
	No of results	16	16	16	16	16	16	16			
	Mean	2378.8	2.62	51.9	245.9	124.8	47.5	1104.2			
	Sample s.d.	469.9	1.21	5.75	202.1	42.8	42.4	697.6			
	't'										
	95% UCL										
	95% LCL										

Table A9.4 *Chemical contaminant testing summary – biopile: 26 weeks after second sludge addition and 19 weeks after end of trials treatment (27.8.98)*

Sample number	Description	Mineral oils (mg/kg)	Monohyd. phenols (mg/kg)	Naptha-lene (mg/kg)	Acenap-thene (mg/kg)	Pyrene (mg/kg)	Benzo-a-pyrene (mg/kg)	EPA 16 PAH (mg/kg)	Mercury (mg/kg)	Lead (mg/kg)	Arsenic (mg/kg)
GML/10/17	Biopile P1	3500	2.38	97.3	1260	211	158	3840			
GML/10/18	Biopile P1	8260	2.72	525	2600	200	124	7200			
GML/10/19	Biopile Q1	3920	4.25	101	1070	160	102	3520			
GML/10/20	Biopile Q1	5480	4.23	298	1730	182	92.2	5330			
GML/10/21	Biopile P2	6200	15.8	184	801	91.8	37.3	2430			
GML/10/22	Biopile P2	6770	11.2	533	1880	165	53.2	5790			
GML/10/23	Biopile Q2	4950	1.79	105	912	126	39	2790			
GML/10/24	Biopile Q2	7450	4.58	387	2400	198	52.8	6990			
GML/10/25	Biopile P3	5700	3.69	263	1700	205	54.4	5230			
GML/10/26	Biopile P3	6610	3.51	551	3670	329	85	10800			
GML/10/27	Biopile Q3	6230	2.71	151	1390	143	41.4	3860			
GML/10/28	Biopile Q3	7230	3.12	215	1230	104	27.2	3330			
GML/10/29	Biopile P4	6490	3.43	411	4160	356	83.5	10700			
GML/10/30	Biopile P4	6460	4.36	280	1560	125	34	4440			
GML/10/31	Biopile Q4	7600	1.66	296	2590	205	47.7	6570			
GML/10/32	Biopile Q4	5530	2.89	131	1850	133	33.1	4020			
	No of results	16	16	16	16	16	16	16			
	Mean	6153.1	4.52	283	1925.2	183.4	66.55	5427.5			
	Sample s.d.	1281.7	3.71	158.1	953.1	72.9	37.5	2525.1			
	't'										
	95% UCL										
	95% LCL										

Table A9.5 *Chemical contaminant testing summary – treatment bed:*
71 weeks after second sludge addition and 64 weeks after end of trials (8.7.99)

Sample number	Description	Mineral oils (mg/kg)	Monohyd. phenols (mg/kg)	Naptha-lene (mg/kg)	Acenap-thene (mg/kg)	Pyrene (mg/kg)	Benzo-a-pyrene (mg/kg)	EPA 16 PAH (mg/kg)	Mercury (mg/kg)	Lead (mg/kg)	Arsenic (mg/kg)
GML/11/01	Treatment bed A1	2100	<1.50	<0.50	<0.50	5.22	2.85	46.9			
GML/11/02	Treatment bed A1	2080	1.67	5.22	10.2	161	107	1020			
GML/11/03	Treatment bed B1	2040	<1.50	23.4	557	209	30.2	1480			
GML/11/04	Treatment bed B1	1270	<1.50	<0.50	<0.50	14.7	6.03	105			
GML/11/05	Treatment bed A2	1680	1.57	10.4	39.3	50.9	19.3	344			
GML/11/06	Treatment bed A2	2020	<1.50	15	9.97	105	27.3	566			
GML/11/07	Treatment bed B2	1200	<1.50	12.2	23.3	22.1	21.7	240			
GML/11/08	Treatment bed B2	1840	1.85	17.4	27.7	26.9	19.4	295			
GML/11/09	Treatment bed A3	2320	<1.5	11	<0.50	<0.50	19.4	563			
GML/11/10	Treatment bed A3	1610	1.55	11.9	39.9	24.2	13.4	245			
GML/10/11	Treatment bed B3	1890	1.55	13.9	27.1	40.9	19.3	339			
GML/11/12	Treatment bed B3	1540	<1.50	9.05	17.1	30.2	17.4	223			
GML/11/13	Treatment bed A4	2450	1.50	17.5	<0.50	<0.50	20.6	866			
GML/11/14	Treatment bed A4	1890	<1.50	12.1	60.4	155	21.9	500			
GML/11/15	Treatment bed B4	2030	1.58	22.3	130	84.8	26.6	712			
GML/11/16	Treatment bed B4	1778	<1.50	1.43	<0.50	5.88	1.43	45			
	No of results	16	16	16	16	16	16	16			
	Mean	1858.8	1.6	11.5	59	58.6	23.4	474.4			
	Sample s.d.	340.8	0.1	7.0	136.8	65.4	23.8	389.7			
	't'										
	95% UCL										
	95% LCL										

Table A9.6 *Chemical contaminant testing summary – biopile:*
71 weeks after second sludge addition and 64 weeks after end of trials (8.7.99)

Sample number	Description	Mineral oils (mg/kg)	Monohyd. phenols (mg/kg)	Naptha-lene (mg/kg)	Acenap-thene (mg/kg)	Pyrene (mg/kg)	Benzo-a-pyrene (mg/kg)	EPA 16 PAH (mg/kg)	Mercury (mg/kg)	Lead (mg/kg)	Arsenic (mg/kg)
GML/11/17	Biopile P1	2590	2.78	24.5	167	144	31.5	965			
GML/11/18	Biopile P1	1500	1.64	12.3	29.7	14.4	14.5	204			
GML/11/19	Biopile Q1	3460	<1.50	10.3	43.5	32.2	17.1	327			
GML/11/20	Biopile Q1	3590	2.09	11.8	<0.50	111	32.7	662			
	No of results	4	4	4	4	4	4	4			
	Mean	2785.0	2.17	14.7	80.1	75.4	24.0	539.5			
	Sample s.d.	964.9	1.58	6.6	73.4	62.1	9.5	343.4			
	't'										
	95% UCL										
	95% LCL										

References

ENVIRONMENT AGENCY (1997)
Interim guidance on the disposal of contaminated soils (2nd edition)
Environment Agency, Bristol

ENVIRONMENT AGENCY (1998)
Waste management licence applications relating to the remediation of contaminated sites, Directorate Instruction No 4/98
Operations Directorate, Environment Agency, Bristol

ENVIRONMENTAL RESOURCES LIMITED (1987)
Problems arising from the redevelopment of gas works and similar sites (2nd edition)
Department of the Environment, HMSO, London

FERGUSON, C AND DENNER, J (1994)
"Developing guideline (trigger) values for contaminated soil: underlying risk analysis and risk management concepts"
Land Contamination and Reclamation, Vol 2 No 3, pp 117–124

FERGUSON, C AND MARSH, J (1993)
"Assessing human health risks from ingestion of contaminated soil"
Land Contamination and Reclamation, Vol 1 No 4, pp 177–186

HARRIS, M R (1996)
Framework protocol for reporting the demonstration of land remediation technologies
Project Report 34, Construction Industry Research and Information Association, London

HARRIS, M R, HERBERT, S M, SMITH, M A AND MYLREA, K (1997)
Remedial treatment for contaminated land, Volume XII: policy and legislation
Special Publication 112, Construction Industry Research and Information Association, London

HEALTH AND SAFETY COMMISSION (1995)
Managing construction for health and safety. The Construction (Design and Management) Regulations 1994 approved code of practice
HSE Books, Sudbury

INSTITUTION OF CIVIL ENGINEERS (1995)
The engineering and construction contract (New engineering contract, 2nd edition)
Thomas Telford, London

LEJEUNE, G, ROSEVAAR, A AND FOX, A (1996)
Pollution potential of contaminated sites – validation of leaching test
AEA Technology, R&D Technical Report P10, Environment Agency, Bristol

MARTIN, I AND BARDOS, P (1995)
A review of full scale treatment technologies for the remediation of contaminated soil,
Report for the Royal Commission on Environmental Pollution
EPP Publications, Richmond, Surrey

MCCAFFERTY, A (1993)
"Derelict land grant policy"
Land Contamination and Reclamation, Vol 1 No 2, pp 73–76

NATIONAL RIVERS AUTHORITY (1994)
Leaching test for assessment of contaminated land, R&D Note 301
Interim NRA Guidance

NATURAL ENVIRONMENT RESEARCH COUNCIL (1967)
Geology of the country around Chesterfield, Matlock and Mansfield
Geological Survey of Great Britain, HMSO, London

PEGG, M AND ASHWORTH, W (1986)
The history of the British coal industry, 1946–1982, Volume 4: The nationalised industry
Clarendon, Oxford

PERRY, J G, THOMPSON, P A AND WRIGHT, M (1982)
Target and cost-reimbursable construction contracts. Part B: Management and financial implications
Report R85, Construction Industry Research and Information Association, London

PRIVETT, K O, MATTHEWS, S C AND HODGES, R A (1996)
Barriers, liners and cover systems for containment and control of land contamination
Special Publication 124, Construction Industry Research and Information Association, London

QUINT, M AND LIMAGE, A (1994)
"Risk assessment methodology at Glory Hole, Portsmouth"
Contaminated land: from liability to asset, IWEM Conference, 7–8 February 1994

SAX, N I (1979)
Dangerous properties of industrial materials (5th edition)
Van Nostrand Rheinhold Company, New York

STOREY, P (1998)
"Grassmoor Lagoons bioremediation field trials"
Contaminated land: current practice in remedial technologies
CIRIA Conference, 24 November

UNITED STATES ENVIRONMENT PROTECTION AGENCY (1995a)
Remediation technologies screening matrix and reference guide (2nd edition)
PB95-104782, United States Environmental Protection Agency, Cincinnati, OH

UNITED STATES ENVIRONMENT PROTECTION AGENCY (1995b)
Abstracts of remediation case studies
EPA 542-R-95-001, United States Environmental Protection Agency, Cincinnati, OH

UNITED STATES ENVIRONMENT PROTECTION AGENCY (1996a)
"Prepared bed land treatment effective in remediating wood preserving wastes at Libby Site"
Bioremediation in the Field Newsletter
EPA 540-N-96-500, United States Environmental Protection Agency, Cincinnati, OH

UNITED STATES ENVIRONMENT PROTECTION AGENCY (1996b)
Bibliography for innovative site clean-up technologies
EPA 542-B-96-003, United States Environmental Protection Agency, Cincinnati, OH

WESTCOTT, F J (1997)
"The drawdown of Grassmoor Lagoons: direct discharge to foul sewer as a cost effective approach to groundwater and surface water remediation"
Land Contamination and Reclamation, Vol 6 No 3, pp 131–142